Cattle Kingdom in the Ohio Valley

1783-1860

PAUL C. HENLEIN

Cattle Kingdom
in the
Ohio Valley
1783-1860

University of Kentucky Press
LEXINGTON

COPYRIGHT © 1959 BY THE UNIVERSITY OF KENTUCKY PRESS
COMPOSED AND PRINTED AT THE UNIVERSITY OF KENTUCKY
LIBRARY OF CONGRESS CATALOG CARD NUMBER: 59-10278

*The publication of this book is made possible partly because of a grant from
The Ford Foundation.*

Preface

THIS IS A STORY OF THE CATTLEMEN IN THE OHIO VALLEY. IT does not concern itself (beyond some mention in the first chapter) with explorers, soldiers, or politicians, as such, although some of the cattlemen were one or another of these. Seven chapters attempt to tell how the settlers occupied the valley, bred cattle, grazed and fed them, drove them to market, and sold them. The first chapter tells of the occupation of the Ohio Valley; the concluding chapter finds the cattlemen looking west again, this time to Missouri and the plains.

I first became interested in the theme of this book through my study of cattle driving from the Ohio country (*Agricultural History* [April, 1954], 28:83-95). When I became aware of the great importance the beef-cattle industry once had in the Ohio Valley, I saw in this larger story an opportunity to study a much-neglected subject. Popular opinion associates the beef-cattle industry with the trans-Mississippi West and seems to assume that this great industry was born full grown; but much of the later business had its antecedent in the Ohio Valley, where also many of the Texas cattlemen received their first introduction into the lore of cattle raising.

The geographical area of my study is arbitrarily limited. The subject is a cattle kingdom bestriding the Ohio River and centering in the Kentucky Bluegrass and the valleys of the Scioto, Miami, Wabash, and Sangamon rivers in Ohio, Indiana, and Illinois. It does not include the Nashville Basin, although this drains into the Ohio; but it does include parts of the present Corn Belt. The upper Ohio Valley was not really a part of this kingdom, for beef cattle had little commercial importance in western Pennsylvania or in the vast stretch of mountainous country between the Big Levels of western Virginia and the Kentucky Bluegrass, although at an early date some high-class cattle were produced around Clarksburg and Kingwood, both in the sources of the Monongahela, and in the bluegrass pastures between Weston and Elkins.

Unfortunately, it was already late to attempt to piece together the story. The documentary remains were dwindling, and sections of the story were passing into tradition or folklore. For example, several years ago the farmhouse of the Kentucky Renicks burned to the ground, engulfing in flames crates of papers of old "Abe" Renick, the important and picturesque cattle breeder. I was told by a friend who used to live in Pickaway County, Ohio, that a collection of manuscripts relating to the cattle industry along Deer Creek was destroyed a few years ago by children who cut them up with scissors to make paper dolls. Although the United States is a young country endowed with a relatively large quantity of manuscript records, time waits for no man, historian or other.

Locating manuscript records is a difficult problem for the agricultural historian. I had the good fortune to meet four persons who had manuscripts, hitherto unexamined, and who took a genuine interest in my project. I wish especially to acknowledge the enthusiasm of these four persons. They are Renick Cunningham, to whom I am indebted for making available to me all the Felix Renick papers in his possession; Joseph I. Vanmeter, to whom I am obliged for scouring his study and finding several Vanmeter and Vause items; Cassius M. Clay, to whom I am grateful for opening to me the voluminous Brutus Clay papers; and Ben Douglas Goff, Sr., to whom I am indebted for letting me borrow the Strauder Goff papers.

Andrew H. Clark, whose inquisitiveness and enthusiasm are contagious, helped conceive the idea of this as a book-length subject.

Vernon Carstensen, my thesis adviser, has brought to the subject his own knowledge of agricultural history, has provided encouragement and ideas, and has gone over the drafts of the manuscript tirelessly and patiently.

Lucien Beckner, who had written "Kentucky's Glamorous Shorthorn Age" (*Filson Club History Quarterly* [January, 1952], 26:37), called upon his intimate knowledge of Kentucky to think up sources.

Edward N. Wentworth shared with me his long experience in livestock history and prepared a criticism of portions of my manuscript.

A research grant from the Graduate School of the University of Wisconsin enabled a year's study of the cattle driving, a study originally proposed by Professor Vernor C. Finch of the geography department.

The librarians of the University of Kentucky manuscript room, the Filson Club, and the Philosophical and Historical Society of Ohio (Cincinnati) have left no stone unturned in summoning the resources of their places to my assistance.

This study is offered as an extension of the work already done by James Westfall Thompson and Charles T. Leavitt. But the study of the beef-cattle industry needs to be filled out with studies of it in the upper South (Virginia, North Carolina, and Tennessee) and after it moves into Illinois, Missouri, and Iowa. This latter project would tie together the work already done by Helen Cavanagh, Paul W. Gates, Clifford Carpenter, John Ashton, Edward N. Wentworth, and others.

Paul C. Henlein

Contents

	Preface	page v
Chapter 1.	Establishing the Industry	1
2.	Early Breeding Practices	21
3.	The Cattle Kingdom, 1834-1860	41
4.	Importations, 1832-1857	74
5.	The Drives over the Mountains	103
6.	Stockyards and Slaughterhouses	130
7.	The Industry Moves Westward	169
	Bibliographical Note	183
	Index	185

Chapter 1

Establishing the Industry

THE OHIO VALLEY BEGINS, TECHNICALLY, AT THE WATER DIVIDE in the Appalachian Mountains. But the range after range of mountains tumbling away to the west blocked off a clear picture of the great valley. The awareness of that valley—the land promised by the brochures of the land companies and later glorified by Thomas Hart Benton as the "Garden of the World"—came to different men at different points as they approached the valley. It came to Daniel Boone and John Finley in 1769 as they stood on Pilot Knob, an escarpment rising 800 feet over the undulating Bluegrass Basin; to many of the Virginians and most of the Carolinians when they reached Big Hill Gap, on Boone's original trace, or drew near to Crab Orchard, on the Wilderness Road; to George Renick in 1793 as he rode along the Great Kanawha down past Charleston and suddenly saw the steep hills fold back from the riverbank and reveal the valley. To the Pennsylvanians who were emigrating overland and to a sprinkling of South Branch Virginians who had struck out for Wheeling and Zanesville, it occurred at some nameless hill on Zane's Trace between Lancaster and the present Kinnickinnick, a nameless hill where the emigrant caught his first glimpse of the gently rolling landscape of prime soil leading to the wide floodplain of the Scioto; and as a back-backdrop, Horseback Knob, the last outrider of the Alleghenies.

As Boone and Finley gazed westward from Pilot Knob, they probably were unaware of the significant soil boundary concealed beneath the forest canopy: that between the shale-based soils of the eastern Kentucky mountains and the limestone-based soils of the Kentucky Bluegrass. The latter, with the limestone soils of the Scioto and Miami valleys, the "Tipton Till" of Indiana north of Indianapolis, and the deep-loess prairie soil of Illinois, formed the soil setting for the cattle-feeding industry in the Ohio Valley.

After the traders, soldiers, adventurers, and speculators, came the first settlers of the Ohio Valley. They belonged to two general groups: natives of the New England and Middle Atlantic states, and natives of the South. The former group, coming by way of the Great Lakes, the Mohawk Valley, the northernmost tier of Pennsylvania counties, Zane's Trace, and the Ohio River, occupied the bottoms along the lower courses of the Muskingum and Hocking rivers, shared the upper valleys of these rivers with the Southerners, got more than a foothold through most of the Miami Valley, and occupied central Indiana. The Southerners, coming through Cumberland Gap, or by way of the Kanawha or the Cumberland Road to the Ohio River and thence to Limestone, or by way of Zane's Trace, filled most of Kentucky, occupied the hill country around the Muskingum and Hocking rivers, penetrated northward into Licking County, occupied the middle and lower Scioto, shared the Miami Valley with the Yankees, occupied southern Indiana, and planted the first non-French settlements in Illinois.

Of these Southerners, a disproportionate share of those destined to influence the cattle industry of the Ohio Valley came from two valleys in what is now West Virginia: the Big Levels and the South Branch. The Big Levels, a pocket in the Appalachian Mountains just west of the water divide, lie around Lewisburg and are traversed by the south-flowing Greenbrier River, a tributary of the New; here, from about 1760, was the ancestral seat of the Renick family of Kentucky. The South Branch of the Potomac flows northeastward the length of what is now the northeastern panhandle of West Virginia; here was the historic cattle country which was the ancestral seat of the Kentucky cattlemen Patton, Gay, and Sanders, and of the Ohio Renicks. Today, Lewisburg in the Greenbrier Valley and Moorefield on the South Branch are the most typically Southern cities in West Virginia. These were the cattle centers of western Virginia, for cattle feeding around Clarksburg and Kingwood was insignificant.

The men from the Big Levels and the South Branch were by outlook and culture lowland Southerners. Many of them, such as Thomas Goff and the Renicks, had little formal education; but if Felix Renick was typical, they were men of considerable intellectual accomplishment—he, at least, read Shakespeare and Addison, mastered surveying, and mapped Indian mounds.[1]

Yankees, Middle Staters, Southerners, foreigners—these were the people who came to the "open and undespoiled" land of the Ohio Valley. Between 1790 and 1800 the settled areas of Tennessee and central Kentucky had joined; the connection was effected by the gradual settlement during the decade of southern Kentucky's "barrens." The Kentucky-Tennessee island of settlement had an extension going north to Cincinnati. The frontier had also extended tonguelike down the Ohio River from Wheeling to a point below Marietta. Settlers had occupied the middle Scioto Valley and had pushed the frontier line up the Miami Valley to the Greenville Treaty Line west of the Great Miami. The next decade witnessed the occupation of southern and central Indiana as fast as Territorial Governor William Henry Harrison could extinguish the Indian title. By about 1830, when the Germans and Irish commenced coming in numbers, the Kentucky mountains, the Pennyrile, and Jackson's Purchase had been settled; and in fact all of the Ohio Valley was now occupied except part of the Indiana and Illinois prairies.[2]

In the rush to occupy the "open and undespoiled" land, the concept of self-sufficiency had considerable force, but it was always subordinate to the desire to produce a readily salable surplus.[3] Self-sufficiency had two counts against it in the Ohio Valley: The climate and soil combined to encourage men to raise corn, a somewhat unsalable surplus crop that could be marketed more profitably in the form of beef and pork; and the Industrial Revolution specialized labor and created a desire for more goods. A corn-and-livestock economy soon gained domi-

[1] Charles S. Plumb, "Felix Renick, Pioneer," *Ohio Archaeological and Historical Quarterly* (January, 1924), 33:16-17.
[2] U. S. Bureau of the Census, *Statistical Atlas, Twelfth Census of the United States* (Washington, 1901), pl. 3.
[3] Rodney C. Loehr, "Self-Sufficiency on the Farm," *Agricultural History* (April, 1952), 26:40-41.

nance. Many of the livestock were hogs, but cattle offered a chance for bigger profits. The beef-cattle business spread gradually to span the valley by 1830.

Although some farmers of today in the Corn Belt raise their own feeder cattle, a much more common practice is to obtain thin stock from the ranges of the Great Plains.[4] In this way the natural resources of land and climate are used to the best advantage and agriculture has been specialized according to function. Both of these developments—utilization of resources and specialization of agriculture—had already occurred by 1830 within the Ohio Valley itself, where cattlemen were making a beef empire of that region. As they learned to utilize the resources of land and water, the Ohio Valley cattlemen classified or sorted out the lands for different agricultural purposes.

But this sorting out of the lands for utilization of natural resources was not alone enough to draw a pattern of range lands and feeding regions. Further considerations were the abundance of corn and the need to market it in a profitable way. In addition to the corn already in the Ohio Valley, the Yankees had brought their yellow "flint" corn, the Southerners their rough white "gourdseed" species, and by 1824 the better drained of the lowlands were laden with corn. But it was selling for only ten cents a bushel.[5] The Erie Canal, which opened next year, was too far from the Ohio Valley to be helpful. The only solution was to convert the corn into marketable produce—whisky, fat cattle, and fat hogs. Thus a shifting grazing and feeding pattern emerged.

As early as 1784 cattle were being driven from the South Branch of the Potomac to the Glades of what is now Kentucky for summer pasture. Cattle trade between Kentucky and the

[4] The terms *stockers, stores,* and, less frequently, *feeders* were used interchangeably to mean thin cattle. Today, stockyards and some farmers make a distinction: by *stockers* they mean cattle so thin that they must be built up more by cheap feeds; by *feeders,* cattle ready for immediate feedlot finishing. The term *stores* is becoming archaic.

[5] Richard Lyle Power, *Planting Corn Belt Culture* (Indianapolis, 1953), 137; W. F. Gephart, "Transportation and Industrial Development in the Middle West," *Columbia University Studies in History, Economics, and Public Law,* 34:84.

South Branch is indicated as late as 1817 by Lewis Sanders' assertion that he included Longhorns in his 1817 importation partly to please "some old South Branch feeders." Probably, too, the South Branch feeders had been getting stock cattle from the Scioto Valley of Ohio for some years before 1811. Central Kentucky cattle were being driven by herdsmen about 180 miles north to the west-central Ohio upland for pasturing. By 1800 this upland, particularly the Darby Plains, was also being used as pasture by Scioto Valley cattlemen.[6]

The first feeding region in the Ohio country was the Kentucky Bluegrass, where the cattle industry developed in the 1790's. Here the hunters and "adventurers" of Boonesboro— Tom Shelby I and Nathaniel Hart—received large land grants that became cattle estates. The log cabins were usually located near springs; the cattle ate the "switch cane" which grew wild among the giant trees.[7] In 1792 Thomas Goff made a trip back to Virginia to fetch a wife; while on this trip, says one local Kentucky story, he saw his horse eating a strange grass (bluegrass) in the Powell Valley and brought some seed back with him to Kentucky.[8] Bluegrass will grow in woods, but it does best after the trees are cleared. Wrote one Kentucky cattleman: "Woodland pastures will keep young stock growing, and old ones on foot, but will not fatten them. A three-year-old Durham, or a five-year-old Patton, and common ox, will get 'stall-

[6] James Westfall Thompson, *A History of Livestock Raising in the United States, 1607-1860*, U. S. Dept. of Agriculture, Agricultural History series, No. 5 (November, 1942), 92; Lewis Sanders to Edwin J. Bedford, May 2, 1853, University of Kentucky Library; Robert Leslie Jones, "The Beef Cattle Industry in Ohio Prior to the Civil War," *Ohio Historical Quarterly* (April, July, 1955), 64:174, 176; Jervis Cutler, *A Topographical Description of the State of Ohio, Indiana Territory and Louisiana* (Boston, 1812), 38-39.

[7] Ben Douglas Goff, Sr., interview with the author, Winchester, Kentucky, November 8, 1955. The cane, three to twelve feet high, had joints along the stalk, from which willowlike leaves grew. John Filson, *Kentucke*, ed. by Willard R. Jillson (Louisville, 1930), 24.

[8] Goff interview. Bluegrass was native in the northern part of both hemispheres, but it would not grow well south of Tennessee. Liberty Hyde Bailey, *Cyclopedia of Farm Crops* (New York, 1922), 373. Apparently it was not common in North America. Edward N. Wentworth writes, "I have always heard that it reached Virginia by chance as some extra feed for livestock, presumably horses, which were imported there." Wentworth to the author, June 19, 1956.

fat' in a year on *open* blue grass."⁹ But in trying to open up pastures, the cattlemen would spare an occasional giant oak for shade for the cattle. If stock are kept off bluegrass in the spring till it grows up well, bluegrass pasture will last the whole year except where the winters are severe.

Bluegrass pasture became one leg, and corn the other, upon which the cattle industry of central Kentucky stood. The corn was shocked in the field following the Virginia custom, with about sixteen hills of corn making one shock. Some of the farmers learned to "riddle"; that is, they would cut only the corn that was ripe and build up the shock gradually. The cattlemen built no barns. They wintered their two-year-olds out of doors on shocked corn, put them on bluegrass in the spring and summer, and then stuffed them with corn until February, when the drive to market began.[10]

A more efficient method of fattening was soiling, or stall feeding, which meant that the animals were actually kept in stalls in winter, or else were allowed to roam about a feedlot covered with straw, as was frequently done in Kentucky and the Scioto Valley. Cattle in a Kentucky feedlot were not as likely to have sheds for protection as were cattle in a Scioto feedlot. Soiling allowed more intensive use of the land than if it were all in pasture; that is, the land would support the fattening of more cattle. The cattle fattened faster, since they wasted no time searching about the fields for herbivorous matter that suited their fancy; and they were less subject to accidents. The stalls, like the more rudimentary sheds, provided shelter; ideally the stalls had walls, ventilation, and dry litter. Feed was provided soon after daylight, at noon, and before sunset. Each steer was served "as much as he can fairly eat with a relish, but the moment he begins to toss it about, . . . it should be instantly removed."[11]

[9] Cassius M. Clay, quoted in *Ohio Farmer* (Cleveland), May 24, 1856.
[10] *Franklin Farmer* (Frankfort, Kentucky), September 23, 1837; Goff interview.
[11] *Scioto Gazette* (Chillicothe, Ohio), October 27, 1849; *Franklin Farmer*, February 10, 1838; Jesse Buel, *The Farmer's Companion; or, Essays on the Principles and Practice of American Husbandry* (Boston, 1839), 249-51.

Some cattlemen fed straight corn in the stalls, but much of this corn passed whole through the animal. Some claimed that corn would be better digested and would "add very much to their fattening" if ground fine and added to the cattle's drink. Such a corn diet, however, failed to supply carotene and Vitamin A. Cattle, moreover, being herbivorous, like to jaw and ruminate grass and plant roughage, and the grass was thought to supply a "bitter" helpful to digestion. The wiser cattlemen, therefore, included shucks, corn stalks, and wheat straw in the winter feed. A few cattlemen tried to reduce the corn ration and to keep the cattle mainly on grass or straw: "I would not during the winter give more than one barrel of corn per head; and another in the month of June on grass—this is the only [way] now [1834] to make money out of cattle or Butchers," wrote William Hart. The butchers, however, could detect grass-fattened cattle by their flabby flesh.[12]

Improved breeds could be ready for market at three and a half years of age, others at four and a half to five.[13] Thus the grazing and feeding operations were combined in the Kentucky Bluegrass. Graziers could rent an occasional large absentee holding that might be as large as 1,400 acres. Stock cattle were available within the Inner Bluegrass itself or might be brought in from outside. A typical farm might have thirty to fifty cattle, twenty-five horses, numerous hogs, some sheep, and a still.[14]

Among the breeders and feeders represented in Fayette County by 1819 were Henry Clay, the Harts, Stephen Fisher, Robert Crockett, W. T. Banton, D. Harrison, John Spears, E. Warfield, and William Smith.[15] In Madison County, Green

[12] *Kentucky Advertiser* (Winchester), October 26, 1816; G. Bohstedt, "Achieve through Correct Feeding," *Shorthorn World* (August, 1952), 37:185-86; *Franklin Farmer*, January 4, 1840; William P. Hart to Virginia Shelby, August 29, 1834, Filson Club, Louisville, Kentucky.
[13] Adam Beatty, "On Grazing and Feeding Cattle in Kentucky," *Southern Agriculture* (Maysville, Kentucky, 1830), 264-72, cited in Elizabeth Ritter Clotfelter, "The Agricultural History of Bourbon County, Kentucky, Prior to 1900" (thesis, University of Kentucky, 1953), 23.
[14] *Western Citizen* (Paris, Kentucky), February 27, 1830, February 19, September 17, 1831; Felix Renick Account Book, in possession of Renick Cunningham, Chillicothe, Ohio.
[15] *American Farmer* (Baltimore), August 25, 1820.

Clay, the uncle of Henry Clay, grazed cattle on the estate which had been given him in payment for being a surveyor with Daniel Boone; at his death in 1828 Green Clay had 114 steers and spayed heifers averaging one year of age, and twenty head of yearlings and calves.[16] The cattle dealer appeared early in the person of Thomas Goff, and the agent-drover in the persons of Lewis Heath of Bourbon County and, later, B. F. Cloud of Clark.[17] Thomas Goff made his farm on Strode's Creek in Clark County and soon added acreage that extended to Hancock's Branch, not far from the old Matthew Patton farmstead.

The second great feeding region to develop in the Ohio country by 1830 was the middle Scioto Valley, lying in Ross, Pickaway, and Franklin counties. Any beeves fattened before 1803 were kept for local consumption, and the only cattle driven out to the South Branch were stockers. By 1803 the valley, which three years before had had to import grain from Pittsburgh, had a corn surplus. The next year (1804), George Renick began fattening sixty-eight head of cattle, which he drove to Baltimore in 1805.[18] Although in that year a destructive freshet occurred on the Scioto and its tributaries, such flash floods were not unusual in the Scioto Valley and there is no evidence that they discouraged corn raising and cattle feeding.[19] In 1810, it is said, 700 cattle were fattened; in that year, Felix Renick commenced bringing in stockers from outside the valley. By 1815, it is reported, some 1,500 cattle were being

[16] Inventory in Brutus J. Clay Stock Book, 1830, in possession of Cassius M. Clay, Paris, Kentucky.
[17] Goff interview; Felix Renick Account Book; James Gay to Strauder Goff, April 28, 1848, in possession of Ben Douglas Goff, Sr., Winchester, Kentucky.
[18] *Scioto Gazette* (Chillicothe, Ohio), November 22, 1831; Jones, "The Beef Cattle Industry," 174; Clement L. Martzolff (ed.), Thomas Rogers, "Reminiscences of a Pioneer," *Ohio Archaeological and Historical Quarterly* (July, 1910), 3:210.
[19] William Renick, *Memoirs, Correspondence, and Reminiscences* (Circleville, Ohio, 1880), 100. Flash floods are known to have occurred in 1836, 1844, 1850. *Scioto Gazette*, April 13, 1836; May 30, 1844; Isaac Van Meter Journal for 1850, in possession of Joseph V. Vanmeter, Chillicothe, Ohio. Robert Leslie Jones fails to prove that because of the flood of 1805 the farmers were afraid to raise corn in the stream bottoms until 1815. Jones, "The Beef Cattle Industry," 186.

fattened. In that year, Felix made a drive to Philadelphia, and thereafter a drive to the East from Ross County occurred as often as every year or two, perhaps oftener. By 1819, Chillicothe had a slaughterhouse. At the same time, herds of fat cattle were being driven out of Pickaway County.[20]

In the late 1820's an estimated 4,000-7,000 cattle were being fattened each year in the Scioto Valley. Some of the local farmers sold their shocked corn to neighboring cattle feeders, who were allowed the privilege of bringing their cattle into lots for feeding, and a few of the feeders got their start by buying up the corn crops of such farmers. By 1832, and probably much earlier, bluegrass had found a place in the feeding regimen; cattle were wintered on corn and grass, grazed during the summer on bluegrass, and stall-fed or lot-fed the following winter.[21]

Although in 1831 the number of stock cattle sent east from the Scioto Valley was two and a half times the number of stall-fed cattle, the unusually large corn crop of that year caused a large number of stockers to be brought in that fall. Thus, in 1832 the Scioto Valley feeders had plenty of fat cattle to send to the eastern markets. George Renick estimated that 12,000 cattle had been fattened in the Scioto Valley in 1831, and that this figure would at least be duplicated in 1832.[22] Felix Renick went to Missouri in the summer of 1832 and drove back to the Scioto Valley a herd of 220 Missouri stockers, of which his nephew William of Pickaway County bought eight head for $120.[23] Felix and his partners may have fed some of the Missouri stockers on shocked Scioto corn in the fall of 1832 before

[20] *Scioto Gazette*, November 22, 1831; Charles Sumner Plumb, "Felix Renick, Pioneer," 10; Felix Renick Account Book; *Western Star* (Lebanon, Ohio), January 26, 1819; *Springfield Farmer* (Springfield, Ohio), February 13, 1819.
[21] *Scioto Gazette*, November 22, 1831, October 3, 1832; *Supporter and Scioto Gazette* (Chillicothe, Ohio), January 13, 1825; Charles T. Leavitt, "The Meat and Dairy Livestock Industry, 1819-1860" (dissertation, University of Chicago, 1931), 93.
[22] Leavitt, "The Livestock Industry," 86-87; *Scioto Gazette*, November 16, 22, 1831.
[23] Felix Renick Memo Book, Acts of 1830-1833, Renick Cunningham collection.

selling them. The local banks were glad to lend money for the purchase of stockers, for the security (the animals) increased in value every day, and the interest amounted to 10-12 percent of the original loan.[24]

Credit, indeed, was a keystone of the feeding industry. The feeder would borrow money from the Bank of Chillicothe by presenting the bank with a bill of exchange drawn upon the bank of an eastern city where he had sold cattle. Then the feeder would hand the Bank of Chillicothe notes to the Madison County grazier from whom he had purchased stockers. The grazier might use the same notes to pay for Illinois yearlings.[25]

Of the Scioto Valley feeders, some—like Felix Renick and Joseph Vause—had been pioneers, or first settlers; others had come in what James Flint calls the third wave: "a capitalist, who immediately builds a larger barn than the farmer, and then a brick or frame house. . . . He erects better fences, and enlarges the quantity of cultivated land; sows down pasture fields, introduces an improved stock of horses, cattle, sheep. . . . He fattens cattle for the market, and perhaps erects a flour-mill, or a saw-mill, or a distillery. Farmers of this description are frequently partners in the banks; members of the State assembly, or of Congress, or Justices of Peace."[26] The typical Scioto Valley feeder was a businessman concerned with sending cash produce to distant markets. The cattle business is inherently riskier than raising hogs, and some of these Scioto Valley feeders took what were extraordinary risks even for the cattle business. In setting out on horseback for Missouri in 1832, Felix Renick had no way of knowing what price would be asked for stockers in Missouri when he got there, or what price the fattened cattle would fetch a year or two later in Philadelphia or the other eastern markets. Such daring was the stuff of empire building.

The Miami Valley carried on cattle feeding on a smaller scale than the Kentucky Bluegrass or the Scioto Valley. The

[24] Leavitt, "The Livestock Industry," 94.
[25] Rudolf A. Clemen, *The American Livestock and Meat Industry* (New York, 1923), 140. [26] Clemen, 81-82.

occupation of the Miami Valley had proceeded surprisingly slowly; Greene County land was still being cleared in 1820, and there were fifty to seventy-five new farms just hacked out of the wilderness, and thirty-two sawmills to maw the falling forest. Traditionally thought of as a country of small, independent farmers, typified by Jeremiah Morrow, the Miami Valley actually seems to have had a number of large holdings and a force of migratory farm labor, despite the proximity of the Indiana frontier in the early years.[27] As one man could tend only twenty-five or thirty acres of corn, a farm of 300 or 400 acres required hired hands, who were housed in log cabins or in the manor house itself, in a room with its own stairway. These large rural establishments differed, however, in one curious respect from those of the Scioto Valley: cash and credit seem to have been in chronic short supply until the 1850's, despite the fact that the valley contained one of the sounder banks of the state (the Miami Importing Company), and this shortage put a considerable crimp in the cattle business, which requires much capital.[28] Although the farmers were turning from wheat to corn as a more reliable crop, the cattle-fattening industry remained small during the 1820's and 1830's.[29] The annals of the Union Village Shakers, however, record a small sale of cattle as early as 1818, and in 1822 much of the beef in the Cincinnati market was coming from Warren County.[30]

The fourth cattle-feeding region, a belt of corn farm stretching between Indianapolis and the Wabash Valley, developed after the War of 1812. During most of its existence until 1834, however, it suffered from its relative remoteness from markets. Flatboating down the Wabash and White rivers to New Orleans was one outlet; but the eastern markets were somewhat

[27] *Western Star*, October 17, 1818; November 11, 1820.
[28] Moses Steddom Account Book, in possession of the Warren County Historical Society, Lebanon, Ohio.
[29] Leavitt, "The Livestock Industry," 83; Tax Duplicate, Warren County, 1836-1839, in Auditor's Record Room, Warren County Courthouse, Lebanon, Ohio.
[30] Josiah Morrow Scrapbook, in possession of the Philosophical and Historical Society of Ohio, Cincinnati; *Western Star*, April 6, 1822.

inaccessible until the National Road reached Indianapolis in the late 1830's and until the canal was built connecting the Wabash River with Toledo. Certainly, it was hard to compete with the Bluegrass and the Scioto feeders, who had a shorter drive to market. The Wabash Valley region, moreover, was settled by Yankees, who lacked the experience in beef production that the Kentuckians, Virginians, and Shenandoah and Pennsylvania Germans had.

A fifth feeding region developed in the Sangamon Valley of Illinois, just outside the Ohio Valley, during the 1830's. The Southerners, who were a significant if not the major element in the population of Illinois in the early years, raised some corn, as was traditional with them, from the time of their arrival in Illinois. They were scored by Yankees, however, for letting their stock get wormy, for depending on erratic native pasture, for neglecting to raise hay, for neglecting dairying, for wintering livestock on corn blades, and for omitting to erect fences or else putting up "worm" fences.[31] Since the New Englanders in Indiana and Illinois also sometimes omitted to erect barns and left corn in the fields, they could hardly criticize the Southerners for these practices. Solon Robinson no doubt had Illinois more in mind than Ohio or Kentucky when he wrote, "Where the most corn is fed and little else, there I find 'scrub breed' in the highest state of scrubbiness."[32] During the whole period to 1834, Illinois corn fattening was of little more than local significance; the grazing of the native prairie grasses and the production of stock cattle were the chief pursuits of the Illinois cattlemen.

The four feeding regions, particularly the Kentucky Bluegrass and the Wabash Valley region, raised some of their own stockers. But all four regions also brought in stockers from the outside; in fact, George Renick estimated in 1831 that perhaps not more than one-half the cattle fattened in the Scioto Valley had been raised there.[33] The stockers came from six ranges: the Kentucky range, the southern Ohio range (on both

[31] Power, *Planting Corn Belt Culture*, 92-105.
[32] Power, 95. [33] *Scioto Gazette*, November 22, 1831.

sides of the Scioto Valley), the west-central Ohio upland, the artificial grasslands of Darke and Mercer counties, the Grand Prairie of Indiana northwest of Lafayette, and the prairie range of Illinois. In the instance of much (but not all) of the Kentucky and southern Ohio ranges, poor soil limited the availability of rich feed. Much controversy arose over whether these ranges could support blooded cattle; currently, the stockers they supplied were scrubs or, at best, part-bloods (usually grades). Presumably nobody suggested that the farmers of these ranges attempt to keep large numbers of blooded cattle; the idea must have been that these rangeland farmers could keep enough blooded breeding stock to supply outside feeders with stockers of improved blood. William Renick of Pickaway County was among the foremost of those who believed that blooded cattle did not require extra feed and that the ranges could therefore support an improved breed of cattle. Many Ohio Valley farmers tried his theory and found him wrong.[34] Youatt's guidebook for cattlemen said it was "absurd" to bring animals from a better to a poorer soil: "The original stock will deteriorate if neglected and half-starved, and the improved breed will lose ground even more rapidly, and to a far greater extent."[35] The agricultural press reported nearly unanimously that lack of good feed and of good shelter would cause improved breeds to fail.[36] Said one farm paper: "Durham cattle if kept no better than the native stock, are little superior to the latter, except as propagators of a progeny that with good keep may be made perfect at will. In hardihood, we believe them inferior to the common stock."[37] Another agricultural paper added: "Without good feed, short horns become ghosts as well as other cattle; experience shows that not only a *scrub*, but even a Devon will do well on pastures where a Durham would almost

[34] W. Renick, *Memoirs*, 59; Charles T. Leavitt, "Attempts to Improve Cattle Breeds in the United States, 1790-1860," *Agricultural History* (April, 1933), 7:56, 60.
[35] William Youatt, *Cattle* (Philadelphia, 1836), 525-26.
[36] Leavitt, "Cattle Breeds," 60.
[37] *Franklin Farmer*, quoted in *Monthly Genesee Farmer* (Rochester, New York), July, 1839.

starve."[38] Thus it was affirmed that natives and partbloods were the proper cattle for the stocker country of Kentucky and southern Ohio.

The knowledge that blooded stock would not do as well as scrubs on poor feed led to a corollary: "In the purchasing of cattle, whether in a lean or fat state, the farmer should on no account buy beasts out of richer or better grounds [than his]. ... It will, therefore, be advisable to select them from stock feeding in the neighborhood, or from such breeds as are best adapted to the nature and situation of the soil."[39]

The Kentucky range was a sprawling area including such diverse country as the foothills and knobs of Montgomery and Powell counties, the Green River Barrens, parts of the Pennyrile, and the grasslands of Marshall County, beyond the Tennessee River. Ohioans obtained stockers from the Kentucky range and also from the Inner Bluegrass itself. An authority writing recently has made the following, possibly misleading, statement: "Within a few years cheap cattle were obviously no longer readily found immediately south of the Ohio, for in 1810 Thomas Worthington purchased 420 head and Felix Renick at least 100 from the Chickasaws."[40] Actually, stockers were readily found immediately south of the Ohio, but the feeders who bought them complained through the years (as we shall see) that these stockers were not cheap. In 1815 Felix Renick left money with Lewis Heath of Paris, Bourbon County, which was only about seventy-five miles from the Ohio River, to buy cattle for him. Heath spent "60 days purchasing, driving, etc." A herd of 296 head was assembled, the purchase prices averaging $24 per head. A. Miller, George Renick, and William Lewis also sent money to Kentucky for buying cattle. Some of the stockers assembled at Paris for Ohioans came from such Bluegrass cattlemen as Harrison, Crockett, Clay, and Bedford, but others probably came from outside the Inner Bluegrass. In 1816 Felix went into Kentucky himself, via Blue Licks, and

[38] *Monthly Genesee Farmer* (July, 1839), 4:121.
[39] *Farmers' School Book*, quoted in *Franklin Farmer*, October 14, 1837.
[40] Jones, "The Beef Cattle Industry," 181.

brought a herd of stockers back to Ohio. William Renick later recalled that as a young man in the years 1821-1825 he had ranged over all of Kentucky as far south as the Green River Barrens buying feeders individually or in collected lots. Stock cattle were still available in Bourbon County, in the Inner Bluegrass, in the early 1830's.[41]

A second source of feeder cattle was the hill country of southern Ohio, on both sides of the Scioto Valley. William Renick purchased thin cattle here during the years 1821-1825. Unfortunately, no records have been located revealing the volume or details of the cattle business in this country before 1834. The stockers coming from the region east of the Scioto Valley—the so-called "Hocking cattle"—were healthy, hardy, and compact, but too small to benefit fully from lavish feeding. The stockers coming from Highland, Adams, and Brown counties were called "Brush Creek cattle," after a creek running into the Ohio River. The "Brush Creek cattle" had the same good qualities as the "Hocking cattle," were a little larger, and fatted early. The southeastern Ohio farmers kept their native cattle on "short pasture" during summer, and in winter put them on browse and stacked hay.[42]

The third range was the west-central Ohio upland. This tableland between the middle Scioto Valley and the upper Miami Valley extends from the southern part of Fayette County northward to Logan County. Although most of Fayette County was forested, long-grass prairies made the upland attractive as an open range; western Pickaway County had clay-soil white-oak prairies, Madison County had a "buck prairie" and the Darby Plains, and Champaign County had "oak openings," King's Creek Prairie, and Pretty Prairie.

We have mentioned herds from Kentucky being pastured in Madison County, Ohio, in 1812. Shortly thereafter, however,

---

[41] Felix Renick Account Book; W. Renick, *Memoirs*, 55; *Western Citizen*, February 27, 1830; September 17, 1831.

[42] Renick, *Memoirs*, 55; Charles W. Burkett, *History of Ohio Agriculture* (Concord, New Hampshire, 1900), 110-11; Jones, "The Beef Cattle Industry," 289.

feeding had begun in the upland. The farmers of Champaign County, who had good corn lands along the creeks and on some of the prairies, were estimated in 1815 to be receiving $100,000 annually for fat cattle, a sum unlikely to be fetched by grass-fattened cattle alone.[43] E. Johnson of Oldtown claimed in 1816 that his Paint Creek bottom land (in what is now Fayette County) produced 3,000 bushels of corn annually and 115 bushels per acre.[44] Western farmers, however, habitually claimed their land would produce 100 bushels per acre.

Despite these beginnings of corn growing and cattle fattening in the upland, most of the upland remained in 1816 an unsettled wilderness because of marshiness, the Virginia military-bounty system, absentee ownership, land aggrandizement by speculators, and—said the Yankees—the laziness of the Southerners who owned the land. This range, like the Carolina cowpens before it, had some of the same trimmings as the trans-Mississippi "cow country" of half a century later. Even before the cow town of South Charleston was founded, other characteristics of cow country appeared, including even gunsmoke. Absalom Funk, an early cattleman in the upland, was wounded by gunfire while helping his brother Jacob resist arrest and extradition to Kentucky; thereafter the bullet-riddled Funk cabin was known as "Funk's Fort."[45] A traveler commented in 1816 that the typical farmer in Fayette County left girdled trees standing, at the risk of their falling on his wife, children, and cattle, and that "young cattle . . . are often destroyed by wolves."[46] Around 1818, travelers reported that the prairies and pastures between Coshocton and the Miami country were crowded with herds of cattle. In the upland, until the end of the 1820's, herdsmen tended the cattle, in herds of 100

[43] Daniel Drake, *Natural and Statistical View or Picture of Cincinnati and the Miami Country* (Cincinnati, 1815), 54.
[44] David Thomas, *Travels through the Western Country in the Summer of 1816* (New York, 1819), 104. Oldtown (Old Chillicothe) is actually in the Miami Valley, but Johnson's holdings extended east into the upland.
[45] L. H. Everts (comp.), *Illustrated Historical Atlas of Fayette County* (Philadelphia, 1875), 7.
[46] Thomas, *Travels through the Western Country*, 104.

to 300, which were marked with brands. Nobody, so far as we know, called these herdsmen "cowboys"; yet they certainly were more like the "cowboys" of the later trans-Mississippi West than were the "cowboys" of the Carolina cowpens, who were called "cowboys" despite the fact that they tended vegetable patches in their slack time.[47] There was plenty of unsettled open range in the upland, either still reserved to be taken up by scrip, or belonging to land speculator Walter Angus Dunn, or belonging to absentee owners such as one Virginian who owned 3,880 acres in the upland.[48] The fencing of large holdings, however, was under way. A typical "stock farm" in the upland in 1819 had 1,800 acres, of which 1,000 were fenced.[49]

The practice of summer-pasturing Ross County cattle in the upland continued even as the open range was being fenced. In 1831 a Madison Countian at Rosedale offered to pasture 400 or 500 head of Ross County cattle on "pasture well-watered throughout the season with living water."[50]

The native cattle of the west-central Ohio upland were known as "Barren cattle" (because portions of the upland were called "barrens"). These were much larger than the "Brush Creek" stockers, but were loose made, harder to fatten, and much subject to disease. One disease, the bloody murrain or "red water," was so called because of the frequent and small discharges of dark and bloody-colored urine. The traders in "Barren cattle" counted on a loss of 3 to 5 percent a year by the bloody murrain.[51] William Renick recalled: "in May, 1826, I

[47] Thompson, *A History of Livestock Raising*, 93; Jones, "The Beef Cattle Industry," 179; Charles Wayland Towne and Edward Norris Wentworth, *Cattle and Men* (Norman, Oklahoma, 1955), 143.

[48] Walter Angus Dunn, List of Debts owing . . . on the 1st of October, 1827, in the University of Kentucky Library; *Scioto Gazette and Fredonian Chronicle*, November 26, 1819. In 1810 Dunn was selling Virginia Military Reserve warrants in Chillicothe as agent for Thomas Harris of Williamsburg, Virginia. Thomas Harris to Dunn, February 13, 1810, University of Kentucky Library. By 1827 Dunn had twenty-seven upland settlers in debt to him for their land.

[49] *Scioto Gazette and Fredonian Chronicle*, June 11, 1819.

[50] *Scioto Gazette*, April 27, 1831.

[51] *Complete Farrier, or Horse Doctor; Also the Complete Cattle Doctor* (Chicago, 1851), 142; Burkett, *Ohio Agriculture*, 110-11; Renick, *Memoirs*, 56.

bought sixty three head, three miles east of Washington Court House [Fayette County], and a month later, fifty-one head four miles northwest of Mt. Sterling [Fayette County], making one hundred and fourteen head in all, and on the 20th of April, 1827, I had but the even one hundred to sell, fourteen having died, all by the murrain."[52] Various "home cures," such as dosing with Sweet Oil or covering the poor animal with suet, were tried, usually in vain, in order to arrest the progress of this disease. Another disease in the upland was called "trembles" because muscular tremors in the animal indicated that he had been poisoned by eating the white snakeroot or the rayless goldenrod. This disease was said to disappear wherever "the country becomes cleared up and the cattle are kept in pastures of cultivated grasses."[53]

The fourth range was the Grand Prairie of Indiana northwest of Lafayette. Here the toughness of the sod, the sloughs filled with a rank growth of cattails, the swarms of greenhead horseflies, and the distance from markets discouraged corn farmers from entering before 1830. The Grand Prairie was, by the midtwenties, a range supplying thin cattle to feeding regions. By then, young Thomas Atkinson was herding cattle in Benton County and what is now Newton County; these cattle he eventually drove to the feeding region of southeastern Pennsylvania.[54] That he took them to Pennsylvania is significant; for not only were the markets for fat cattle distant, but even the markets for thin cattle. Some of the Grand Prairie's thin cattle presumably were taken by corn farmers of the Wabash Valley; yet this belt produced, as we have mentioned, a significant number of its own stockers. The next nearest market for the Grand Prairie's thin cattle before 1830 was the Scioto Valley; here, despite the interest the local feeders showed in obtaining stockers from the West, the more customary (and,

[52] Renick, *Memoirs*, 56.
[53] *Ohio Cultivator* (Columbus), August 1, 1845.
[54] Paul W. Gates, "Hoosier Cattle Kings," *Indiana Magazine of History* (March, 1948), 44:3, 5; Sherman N. Geary, "The Cattle Industry of Benton County," *Indiana Magazine of History* (March, 1925), 21:28.

heretofore, usually ample) sources were the southern Ohio range and the Kentucky range. The Bluegrass feeders, amply supplied with Kentucky stockers, did not need Indiana thin cattle. So Atkinson took his herd to Pennsylvania.

The fifth range was the great prairie of Illinois, including Champaign and McLean counties, and extending west into the Sangamon Valley and southwest to a point near Covington. Cattle were fond of the coarse, strong "prairie grass," which, however, was susceptible to destruction by feeding or mowing; "yard grass," which appeared on heavily grazed or trodden land, also appealed to cattle, and besides it was drought resistant. Settlers destroyed the prairie flies by burning the prairie grass in June instead of the fall and then using it for pasture after it grew anew.[55]

Stock ran at large on the open prairie range. Around Edwardsville "they turn everything out to run at large and when they want to use a horse or oxen they will have to travel half a dozen miles to them through grass and weeds higher than a man can reach when on horseback and the grass and vines are so rough that nothing but their leather hunting shirts and trowsers will stand any chance at all."[56] Cattle were fenced out, not in. A fence needed to be "bull strong, horse high, and pig tight." Often a farmer would live for many years on his farm without erecting a barn or even a shed for his cattle. All calves were kept, even if inferior and even if they could not be supported properly, because the calves enticed the cows home at night from the open range.[57]

[55] *Scioto Gazette*, October 5, 1843; W. C. Flagg, *Agriculture of Illinois, 1682-1876*, Illinois Department of Agriculture, *Transactions* (1875), 13:311-12. "Nimblewill," a kind of florin grass or running couch grass which sprang up where prairie grass had been destroyed, did not appeal to cattle. Cattle would reluctantly eat "buffalo clover," but only if it was cut for them. White clover, which like red was thought pernicious to livestock, was rare in the prairie; but bunches of it began to appear in 1818 along the Goshen–St. Louis road, followed almost immediately by bluegrass. Bluegrass was highly prized in Illinois as pasture. Timothy, strictly a cultivated grass, was thought by Illinoisans until about 1831 to be unable to hold its own against weeds.

[56] Richard Bardolph, "Illinois Agriculture in Transition, 1820-1870," *Journal of Illinois State Historical Society* (September, 1948), 59:249.

[57] J. M. Peck, *A Guide for Emigrants* (Boston, 1831), 166-67.

Some of these grass-fed cattle were butchered without first going to a corn region such as the Scioto for finishing. Herdsmen drove one herd of 700 north from Springfield in 1825 to the commissary at Fort Howard, Wisconsin Territory. Despite some of the casual methods employed on the prairie, Illinois young cattle on the prairie were said to be ready for the knife by the first of June or even the middle of May, and an Illinois booster claimed that "the fat is well proportioned throughout the carcase, and the meat tender and delicious." These cattle were fat though not large, seldom weighing over 600 or 700 pounds.[58]

The open prairie range of Illinois was, nevertheless, a source of thin cattle for Ohio graziers and feeders as early as 1818, and probably even 1816. David Thomas' reference to Illinois cattle marching through Ohio in 1816 allows the reasonable supposition that some of these cattle were purchased by Ohio graziers and feeders.[59] The year 1816 must be very nearly the exact date for the beginning of a grazing and feeding relationship between Illinois and Ohio; the number of cattle fatted on the Scioto, it will be recalled, had more than doubled between 1810 and 1815, and in the former year Felix Renick had commenced bringing in stockers from outside, though not at first from the West, but from the more familiar country south of the Ohio River.

By 1834 most of the lands of the Ohio Valley had been differentiated or sorted out, and agriculture had become specialized within them. The Scioto Valley now ranked with the Kentucky Bluegrass as a great cattle-fattening region. Despite brief flurries of interest in other crops and types of farming, the corn-and-livestock economy had become firmly seated in these regions.

[58] Eric E. Lampard, "The Rise of the Dairy Industry in Wisconsin: A Study of Agricultural Change in the Midwest, 1820-1920" (dissertation, University of Wisconsin, 1955), 110; Peck, *A Guide for Emigrants*, 166-67; James Hall, *Statistics of the West at the Close of the Year 1836* (Cincinnati, 1836), 146.

[59] Clemen, *The American Livestock and Meat Industry*, 44; Thomas, *Travels through the Western Country*, 120.

Chapter 2

Early Breeding Practices

CATTLE CAME, OF COURSE, WITH THE FIRST SETTLERS. WHAT breed these cattle were depended upon the part of the eastern seaboard that the settler came from. The New Englanders brought Devons ("Rubies"); big, yellow Danish cattle; the short-coupled red-and-white and black-and-white Dutch cattle; cattle of dubious origin which they had obtained from Virginia; and animals that were mixtures of these strains. The Middle Staters brought Holland cattle; coarse, Flemish-type beasts; West Indian cattle; and crossbreeds of these. The Southerners brought Spanish cattle; English cattle; animals derived from an importation—to be discussed later—in 1783; and the inevitable crossbreeds. There is no evidence that an improved beef breed of any of the above, except the 1783 stock, reached the Ohio Valley. Instead, what came were mongrel cattle, with a probable predominance of Devon and Spanish blood.[1] The cattle already in the Ohio Valley when the Americans came were Spanish, from Florida and the Chickasaw country, and small, black Canadian cattle at Kaskaskia.[2] The mongrel cattle from the East, the cattle already in the Ohio Valley, and the mixtures of these two comprised what men in the Ohio Valley called the "native" race or "scrubs."

The scrubs failed to utilize fully the rich feed of the Ohio Valley. Therefore, men turned their attention to developing an improved breed, which could be done by upbreeding the scrubs or by bringing improved breeding stock from Britain.

The "native" cattle or "scrubs" did offer an advantage: though they were not able to utilize rich food fully, they were not fussy about feed or shelter; therefore, if they strayed away, it would not require "a year's labour and produce ... to replace [them]." Presumably this advantage would be lost if the scrubs were subjected to upbreeding. Only New Englanders seriously considered trying to breed up an improved type of native cattle.[3] Some other American farmers as well were aware that upbreeding might be practiced using solely the native cattle with appropriate techniques; this was all that had been done by Bakewell, Colling, and their English contemporaries, or was being done by the developers of the Poland-China hog in the Miami Valley.[4] As a matter of fact, however, all the leading cattlemen of the Ohio Valley thought it more feasible to acquire pedigreed British cattle as breeding stock; that way the United States could, and did, catch up after its late start.

The development of the Kentucky Bluegrass as a cattle-feeding area in the 1790's provided the first impetus for breed improvement in the Ohio Valley. Persons living on the eastern seaboard imported stock from England, and some of this stock was brought into the Ohio Valley almost as soon as Americans began to raise cattle in the valley; and in 1817-1818 a couple of importations were made directly to the Ohio Valley. By 1830 the Scioto Valley feeding area had developed to a degree where it was able to penetrate deeply the eastern cattle markets and there offer new competition to Kentucky. The volume of feeding and the distant marketing of cattle both stimulated breed improvements, and both in turn were stimulated by breed improvement.

The need for more efficient beef animals having been recognized, several factors created a favorable climate of opinion for breed improvement. The widespread introduction of clover, in connection with gypsum, in New England and the Middle Atlantic States from 1785 to 1810 caused farmers to think that even without the rich corn lands of the Ohio Valley they could

[1] Charles Wayland Towne and Edward N. Wentworth, *Cattle and Men* (Norman, Oklahoma, 1955), 135-42.
[2] W. C. Flagg, *Agriculture of Illinois (1682-1876)*, in Illinois Department of Agriculture, *Transactions* (1875), 13:321.
[3] Charles T. Leavitt, "Attempts to Improve Cattle Breeds in the United States, 1790-1860," *Agricultural History* (April, 1933), 7:52-55.
[4] Robert Leslie Jones, "The Beef Cattle Industry in Ohio prior to the Civil War," *Ohio Historical Quarterly* (July, 1955), 64:304; Otis Rice, "Importations of Cattle into Kentucky, 1785-1860," *Register of the Kentucky Historical Society* (January, 1951), 49:41.

support improved breeds. The agricultural press and the agricultural societies promoted interest in upbreeding. Still another factor was the example of the improvement that had been made in fine wool sheep resulting from the importation of Merinos just before the War of 1812. The Middle Atlantic States lagged far behind the Ohio Valley in breed improvement because they were more interested in grains than livestock. New England practiced upbreeding on native stock, which its farmers preferred to Shorthorns.[5]

Several problems beset those Ohio Valley cattlemen who attempted to better their herds. Any animal may be called "purebred" if both its sire and dam are registered. Obviously this gave the English an opportunity, of which some availed themselves, to defraud their American cousins by registering inferior animals and sending them off to America as "purebreds." The genealogy of American breeding cattle was recorded, if at all, in herd books. Some American cattlemen listed certain of their animals in the English Herd Book, which was started in 1822.[6] Others, like Felix Renick and Dr. Samuel Martin, kept their own herd books.[7] The "Kentucky Stock Book" was started about 1837, with Dr. Martin arranging the pedigrees; "it is thought there will be room for all cattle of three quarters [blood] and over."[8] Lewis F. Allen published the initial volume of his *American Herd Book* in 1846 in Buffalo, but there were almost no Ohio Valley entries. His second volume (Buffalo, 1855) was a tremendous accomplishment, for it included about 3,000 pedigrees, among which were pedigrees contributed by all the leading cattle breeders of the Ohio Valley.

Ohio Valley cattlemen were aware that phenotype was based on two elements: appearance (correct conformation and good quality flesh and skin) and performance (the ability to pro-

---

[5] Leavitt, "Cattle Breeds," 52-55.
[6] Alvin H. Sanders, *Shorthorn Cattle* (Chicago, 1900), 282-86.
[7] Dr. Martin's herd book has survived and may be found in the library of the University of Kentucky.
[8] *Franklin Farmer* (Frankfort, Kentucky), October 14, 1837.

duce a large amount of flesh per unit of feed). Despite the fact that the latter was the basic reason for upbreeding, the study of performance suffered because of the difficulty in testing and comparing findings and because of the emphasis the livestock shows placed on appearance, which was considered indicative of performance. Ohio Valley cattlemen have used the grading system; a purebred bull is mated to a scrub dam, and the offspring ("grades"), if heifers, will be mated to a purebred bull.

The cattlemen could consult a breeders' guidebook such as Youatt's American edition (1836). In view of the connections between the Ohio Valley cattle industry and Philadelphia, it may be assumed that this work, published in Philadelphia, was known to at least a fair proportion of Ohio Valley breeders; a volume reached Paddy's Run, in the Miami Valley. Youatt claimed that, other factors being equal, offspring were more influenced by the sire than by the dam; but that the offspring of a half-blooded bull and a full-blooded cow would be more influenced by the dam, because the bull's good traits did not extend far enough back in his ancestry.[9] This was an optimistic view for the Ohio Valley, for it meant that a full-blooded bull not only had more offspring than a cow, but also that he had more influence over the offspring. Therefore, much could be accomplished by bringing in a few full-blooded bulls. English Shorthorn breeders at this time were putting year-old heifers to the bull, although Youatt advised waiting until the animal was two to two and a half years of age, or even three in the case of a bull. He warned against inbreeding, saying that a new bull should be introduced into the herd every third year; this advice, however, as we shall see, was disregarded by one of the most successful of the Ohio Valley breeders.[10]

Our narrative may now proceed to a chronology of early breeding developments in the Ohio Valley, culminating in the "Seventeens" controversy. This controversy reveals the full

[9] William Youatt, *Cattle* (Philadelphia, 1836), 523-27.
[10] *Scioto Gazette and Independent Whig* (Chillicothe, Ohio), September 10, 1834; Youatt, *Cattle*, 523-27.

force of one of the problems, already mentioned, of early breeding—the problem of records and their importance.

Even before farmers had a chance to occupy any of the Ohio Valley, an importation of British breeding cattle to the United States was made by the partnership of Gough and Miller, a Virginian and a Marylander respectively, as soon as peace was signed in 1783. Some of these cattle, and some cattle of later importations by the same men, reached the South Branch of the Potomac before 1790. There they were husbanded by the Patton family. While other emigrants to the Ohio Valley were taking mongrel cattle with them, the Pattons took animals of this fresh strain. The two Patton boys, John and James, left the South Branch farm of their father, Matthew Patton, and with their brother-in-law, James Gay I, migrated to Kentucky (near Winchester), bringing with them a bull of uncertain ancestry and some grade heifer calves not over a year old. In 1790 Matthew Patton followed. Sanders says that he brought with him the full-blooded bull Mars and the full-blooded cow Venus. Matthew Patton's grandson, Benjamin Harrison, disagreed, saying that Matthew did not buy Mars and Venus until 1795, five years after he had migrated to Kentucky; according to Harrison, the cattle the elderly man brought with him were a "long-horned" bull and six heifer calves sired by this bull.[11]

In any event, Mars and Venus did reach Kentucky sooner or later; these were Gough-Miller cattle, or their descendants, and as such, had no pedigrees (there were few in that day) but were possibly Shorthorns. Venus bred two bull calves out of Mars in Kentucky and died shortly thereafter. One of these calves went to Jessamine County (also in the Bluegrass), the other to Ohio, probably the Scioto Valley. Where the bloodline went after that is unknown. Sanders says that Mars got many calves on the native cows in Kentucky; if so, these calves were, of course, grades. Harrison denies, however, that Mars served many of the native cows of Clark County; Mars was mated

---

[11] A. Sanders, *Shorthorn Cattle*, 164-68; Rice, "Importations," 36-37.

chiefly, he says, to the cows owned by the Pattons and Gay. Others would have had to pay $2.00 per cow for the services of the bull, "which price was considered . . . extravagant." Whether or not Mars was allowed to run for breeders, the half-blooded bull calves sired by him were allowed to do so and were available to all of Clark and part of Bourbon County.[12] These, plus the few descendants of Venus, comprised the beginning of the race known as "Patton cattle."

Because this race of cattle consisted of grades and other part-bloods, much confusion has always existed over which English breed was the dominant strain in them. Gray, a careful scholar, has said that many Kentuckians believed the Patton cattle to be Devons, and a recent authority has accepted Gray's assertion; because, however, the "Pattons" did not have the characteristic ruby-red color of Devons, and because the Kentuckians were in frequent contact with Gough and Miller, the importers, this assertion is hard to believe.[13] Both Allen and Sanders have claimed that the "Pattons" were Shorthorns.[14] More recently, however, James Douglas Gay has stated the belief that Mars, as revealed by his facial marking, was predominantly Hereford.[15] Felix Renick, who entertained many cattlemen at his house and who eventually bought a herd of "Pattons," believed that they were improved Bakewell Longhorns.[16] Checking the evidence years later, Brutus Clay reported that the original Patton stock had the long horns and bad dispositions of Longhorns, but the big teats of Shorthorns.[17]

[12] Rice, 36-37; Elizabeth Ritter Clotfelter, "The Agricultural History of Bourbon County, Kentucky, Prior to 1900," (thesis, University of Kentucky, 1953), 2.
[13] Lewis C. Gray and Esther K. Thompson, *History of Agriculture in the Southern United States* (Washington, 1933), 2:850; Jones, "The Beef Cattle Industry," 305; Rice, "Importations," 37.
[14] Lewis F. Allen, "Improvements of Native Cattle," in "Report of the Commissioner of Agriculture for the Year 1866," *House Executive Documents*, 39 Cong., 2 Sess., 107:296; A. Sanders, *Shorthorn Cattle*, 164-68.
[15] Ben Douglas Goff, Sr., to the author, March 23, 1957.
[16] Jones, "The Beef Cattle Industry," 305. The use of the horns for powder horns indicates (contrary to Jones) that the horns were relatively short.
[17] Brutus J. Clay, "Thoroughbred Cattle," Paris, Kentucky, n.d., MS in possession of Cassius M. Clay, Paris.

Probably the "Pattons" had more Shorthorn blood than anything else, and probably the long horns characterized only a few of the animals, a few that had an unusual infusion of Longhorn blood. It should be remembered, finally, that at least until the 1830's the words "Longhorn" and "Shorthorn" had two meanings: specific breeds, or simply general categories under one or the other of which every breed might be classified.

About 1799 John moved from Kentucky to Ross County, Ohio, taking an unknown number of "Patton cattle" with him. He bought a farm next to the Indian Creek farm of Felix Renick, another newcomer, and was elected to the territorial legislature. Shortly after 1803, he died, and his entire herd of "Pattons" was purchased at the administrator's sale by the Renick brothers, Felix and George. "Patton cattle" are first recorded to have arrived, either from the Renick herd or directly from Kentucky, in the Miami Valley in 1807, when they were introduced by one D. Wilder; thereafter they were crossed with the native cattle of the Miami Valley.[18]

While the Ohio Renicks were acquiring John's herd of "Pattons," some more Miller cattle were beginning to come into the Ohio Valley. Between 1803 and 1811 three bulls were purchased from Miller and brought to Kentucky, where Mars already was. These were Pluto, brought in 1803 by Daniel Harrison, James Patton, and James Gay; Buzzard, brought in 1810 by Captain William Smith; and Shaker, brought by the Shakers. These bulls probably had no native blood whatever in them; that does not necessarily mean, however, that they were pure Shorthorns. Dr. Martin's herd book and Harrison's statement both agree that Buzzard's dam was a long-horned cow that Matthew Patton had sold to Miller before leaving for Kentucky. Pluto, a large red or brindle bull with small head and neck and light, short horns, was bred to grade heifers sired by Mars; about 1812 he was taken to Ohio and died soon thereafter. Buzzard was used in many Kentucky herds, including

[18] William Renick, *Memoirs, Correspondence, and Reminiscences* (Circleville, Ohio, 1880), 16; Charles W. Burkett, *History of Ohio Agriculture* (Concord, New Hampshire, 1900), 107-108.

28 CATTLE KINGDOM IN THE OHIO VALLEY

that of the Pleasant Hill Shakers in Jessamine County. Shaker was used by both the Pleasant Hill and the Union Village, Ohio, Shakers, and perhaps by the Hutchcraft family of Bourbon County. Most of the western progeny of Pluto and Shaker, therefore, were grades, except those whose dams had been sired by Mars. These exceptions we may call three-quarter bloods. Miller sold some more bulls, in adition to these three, to Kentuckians; and a Mr. Inskip brought to Kentucky a bull called the "Inskip Brindle," a mixture of Miller stock and Patton stock left in Virginia by Patton. But the three, Pluto, Buzzard, and Shaker, were the big names in the herd books.[19]

Herefords, meanwhile, became a subject of discussion and shortly made an appearance in Kentucky. The Pattons, in fact, may have used a Hereford cross. In 1815, while on his way back from Ghent, Henry Clay stopped at a cattle show in London. Impressed by the attention the Herefords were receiving, he decided to try them in the Kentucky Bluegrass. Clay did not select any stock at the time, but instead he returned home and carried on a preliminary correspondence with Peter Irving (brother of Washington Irving), who resided in Liverpool. Joseph Smith, one of the best judges of cattle in England, was hired to make the selection. Four Herefords were brought across the Atlantic in March, 1817, on board the *Mohawk*: two bulls, a cow, and a heifer, purchased for £105 sterling. On the drive through Virginia one of the bulls died from eating too much red clover. The surviving animals reached the Winchester area of Kentucky, where they were placed with Isaac Cunningham's Patton cattle. "But," one breeder later recalled, "the Herefords made no favorable impression."[20]

Some opinion in Kentucky lauded the Patton cattle and other descendants of Miller cattle: "The breed of cattle which the Cunninghams, Pattons, and Gays have weigh as much at 2

---

[19] A. Sanders, *Shorthorn Cattle*, 168-69; Rice, "Importations," 37; Martin, Herd Book; B. Clay, "Thoroughbred Cattle."
[20] Clotfelter, "The Agricultural History of Bourbon County," 2; Richard Laverne Troutman, "Henry Clay and his 'Ashland' Estate," *Filson Club History Quarterly* (April, 1956), 30:166; Lewis Sanders to Edwin J. Bedford, May 2, 1853, in University of Kentucky Library.

years old as yours [scrubs] will do at 6—and in some instances more—It would be completely within your power to procure some of their breed—neither the expense nor trouble would be considerable."[21] Brutus Clay, on the other hand, later wrote that the Kentuckians found the Patton stock "difficult to fatten early." Felix Renick apparently got some more "Pattons" or some Miller-based stock in 1816, for in that year he bought 100 steers at $75.00 per head from two Kentuckians. These he brought home to Ohio, culled out twenty-five, and replaced them with the same number of "tops" of his own breeding and feeding.[22] Felix had made his first drive to the east-coast markets the year before, and this year Missouri range cattle were appearing in the eastern markets;[23] Felix probably realized that the Ohio feeders would have to offer more highly bred beef to meet this competition.

The first importation of British breeding cattle direct to the Ohio Valley was made by Lewis Sanders of Bourbon County, Kentucky, in 1817. This importation consisted of four pairs of Shorthorns, one pair of Longhorns, and one pair of Herefords.[24] A contemporary account reported: "The order was sent to the commercial house of Buchanan, Smith, and Co., Liverpool, without limit as to price, ordering a selection of the best young cattle that could be procured for breeders; if the prime, or first cost exceeded thirty guineas each, the order was to have been curtailed to half the number of cattle; but still instructed to procure the best that could be had. They employed a Mr. Etches to go into the counties producing the best cattle, to make purchases." The invoice gives some vague pedigrees:

"A bull bred on the River Tees, got by Mr. Constable's bull, brother to Comet.

A Holderness bull out of a cow that gave 24 quarts of milk per day, large breed.

---

[21] James Morrison to Isaac Shelby, n.d., in possession of the Filson Club, Louisville, Kentucky.

[22] B. Clay, "Thoroughbred Cattle"; Charles Sumner Plumb, "Felix Renick, Pioneer," *Ohio Archaeological and Historical Quarterly* (January, 1924), 33:21.

[23] See Chapter VII.

[24] L. Sanders to Bedford, May 2, 1853; A. Sanders, *Shorthorn Cattle*, 175-76.

A bull from Mr. Reed of Westhorne, by his own old bull.
A Holderness bull, got by Mr. Ware's bull.
A Heifer of the Durham breed.
3 Heifers, bred on the river Tees."

The commercial house sent a letter to Sanders and his agents in America telling of the preparations for the voyage: "At the expense of the ship he [the captain] has erected separate stalls for every beast, with the planks placed so, that in rough weather, they would not be scratched by the ship's rolling." In pleasant weather the animals were occasionally allowed to mingle on the deck. The contemporary account reports that upon landing at Baltimore all the cows were with calf; but whatever became of these calves is unknown. Of the four Shorthorn cows, one died en route in Maryland.[25]

The other three—Mrs. Motte, the Durham Cow, and the Teeswater Cow—came to have numerous descendants, which have always been known as the "Seventeens," in reference to 1817. Mrs. Motte was bred to San Martin (one of the Shorthorn bulls) and produced four red heifers; to Tecumseh (another of the Shorthorn bulls) and produced one heifer and five bull calves. The Durham Cow also proved prolific, her last four calves being sired by her own son, Napoleon, by San Martin.[26] None of these offspring were grades, for they were the produce of inbreeding; their pure-bloodedness can be questioned only if the purity of Sanders' imported Shorthorns themselves is questioned. Of course, the three imported Shorthorn cows did have some calves by other bulls, and they are a different matter.

All the rest of his life Sanders was pestered by people asking about the pedigrees for his importation. In 1853 he reminded fellow cattleman George M. Bedford, the man in large measure responsible for the pedigree craze, that "the year 1816 . . . was . . . some eight years before the commencement of Coates Herd

---

[25] *Cattle: A Collection of Papers, Giving an Account of the English Cattle in Kentucky and Extracts from Various English Publications, Shewing the Value and Importance of the Improved Breed of Cattle* (Lexington, Kentucky, 1817), 6-8; A. Sanders, *Shorthorn Cattle*, 175-76.
[26] A. Sanders, 175-76.

Book." At that time, Sanders explained testily, "A general peace had recently taken place throughout Europe, all kinds of agricultural produce was much cheapened, the agent purchased six pairs, no pedigrees." Sanders thought the agents had made good buys, in view of the high prices blooded cattle had been bringing in England. He vowed that he and his two partners "had no more to do in making the selection than you had." They simply had sent out agents to make the selections.[27]

The descendants of the Sanders importation spread through the cattle-raising regions of Kentucky and even into Ohio. General James Garrard of Bourbon County liked the "Seventeens" and probably had more of them than any other person in the Ohio Valley. Captain William Smith, a partner of Sanders' in the importation, had taken a fancy to the Longhorn pair, and he became instrumental in spreading descendants of these two imported Longhorns (the bull was Rising Sun) through the Bluegrass; for he took them to his farm and put them with a Longhorn bull he already had (Bright), and for a time he had the Durham Cow on the farm to be served by the two Longhorn bulls. Rising Sun's progeny became much in evidence in Clark and Bourbon counties.[28] Gradually—and unhappily for the later reputation of the Sanders importation— the term "Seventeens" came to be applied not only to the pure descendants of the Shorthorn cows but also to this Longhorn line. It is hard to say how much prejudice existed against Longhorns in 1817. George Renick of Ross County, Ohio, heard about Smith's Longhorns and came down to Kentucky to look at them. "Capt. Smith sold the first bull calf from his Longhorn cow to George Renick . . . for one hundred pounds Ky. currency," Lewis Sanders later recalled, "but could have had five hundred dollars if he had asked that sum."[29] George's brother Felix, on the other hand, who already had "Pattons" or Miller-

[27] Sanders to Bedford, May 2, 1853.
[28] Clotfelter, "The Agricultural History of Bourbon County," 2, 3; Lewis F. Allen, *History of the Shorthorn Cattle* (Buffalo, 1872), 163-64; Sanders to Bedford, May 2, 1853.
[29] Sanders to Bedford, May 2, 1853.

based cattle, chose not to bring into his herds any—or any more —Longhorn blood.

George Renick used the Longhorn bull for some years in the Scioto Valley. When thirteen-year-old Brutus J. Clay, later of Bourbon County, Kentucky, rode on horseback down the valley in 1821, he happened to see a herd of cattle that made him take another look: "When . . . taking a horseback journey from Columbus to Circleville, in the vicinity of which latter town the Renick brothers owned large landed estates, we saw a herd of a dozen or more Long-horned cattle grazing in a field by the side of the road. . . . We rode up to the fence, hitched our horse, and went into the field to view them. They had every appearance of being either thorough-bred, or high grades of the Longhorn breed, with long drooping horns, pushing forward beyond their noses, or falling below their jaws, light brindle in color, with white stripes along their backs. . . . They were long-bodied, a little swayed in the back, not very compact in shape, but withal imposing animals to the eye. We made no inquiries about them at the time, as we then knew little of breeds of cattle. Thirty years afterwards being again at Circleville, and having a better knowledge of breeds, on inquiry for cattle of that character, we could find no trace, nor even a recollection of them among the older farmers of the vicinity."[30] Probably these cattle were grade Longhorns. That white stripe down the back was a characteristic of the small, wild Kerry cattle of Ireland. How or when Irish cattle reached America is a mystery, but presumably they must have been brought by some Scotch-Irish immigrants, after which some of them may have strayed and reverted to their wild state. In any event, some strange crosses apparently occurred in the Scioto Valley.

While Longhorn blood was spreading through the Ohio Valley, the offspring of Sanders' imported Shorthorns were growing up. Apparently the best Shorthorn progeny out of the importation was sired by the two Shorthorn bulls San Martin and Tecumseh. An excellent progeny was deriving from the

[30] Allen, *History of the Shorthorn Cattle*, 163-64, 168-70.

three imported Shorthorn cows. The Ohio Shakers took the bull Stonehammer, who had been dropped by Mrs. Motte. By now the three cows had grandchildren. The produce of Sylvia, by San Martin out of Mrs. Motte, included the bull Exchange, born in 1826 and owned by former Governor Allen Trimble of Highland County, Ohio, and the bull Duroc, born in 1828 and owned by the Ohio Renicks. The produce of Lady Durham, by San Martin out of the Durham Cow, included the heifer Lady Macallister, born in 1835 and owned by an Illinoisan, James N. Brown. Besides these three, there were about thirty-seven other female grandchildren of Sanders' three imported Shorthorn cows.[31] The full number of Seventeens, even excluding the Longhorn line, was of course much larger than this.

Judged by their visible merits, the Seventeens included some prize animals, but also some animals which "in the opinion of the best judges and breeders of cattle in . . . [the Scioto] Valley, have [1831] added neither to the size, nor form of the original [Gough-Miller] cross."[32] Obviously, not all of these animals could have been purebred, although after five crosses with scrubs using the grading system they would have been fifteen-sixteenths pure Seventeen. As the 1830's wore on, however, and the year 1817 receded further back into the past, doubt crept in even about those that had been supposed purebred. The basis of this doubt was twofold: the obscure origins of the Sanders importation, and the fact that the Longhorns had been around since 1817. That Sanders had brought back in 1817 four pairs of apparent Shorthorns did not satisfy particularists, who feared that Longhorn blood might "out" in some succeeding generation. In 1835 Alexander Waddle of Clark County, Ohio, questioned the purity of Mrs. Motte.[33]

[31] Allen, *History of the Shorthorn Cattle*, 168-70; Martin Herd Book; L. Sanders to Bedford, May 2, 1853.
[32] *Scioto Gazette*, November 2, 1831.
[33] Lewis F. Allen to Brutus J. Clay, May 19, 1835, Cassius M. Clay collection. Allen himself disagreed with Waddle and called Waddle's questioning "a quibble." "I shall adopt a *liberal* policy [in making my herd book] as you suggest," wrote Allen to Clay, "for the reason that so long a period has elapsed [during which?] Shorthorns have been bred in this country without being publicly recorded."

He claimed that although the English Herd Book was not published until 1822, English breeders had kept records of their stock long before that, and in fact there were pedigrees at least as early as 1777. Lewis F. Allen took up this challenge, made an investigation of these early pedigrees, and put together a full pedigree for Mrs. Motte: "She was got by Adam (717) dam Sterling by a son of Mr. Maynard's yellow cow by Favorite (252) gr. d. by a son of Hubback (319) gr. g. d. by Mansfield (404) gr. gr. g. d. Young Strawberry by Dalton Duke (188) gr. gr. gr. g. d. Lady Maynard, the finest cow ever owned by Charles Colling."[34] When H. H. Hankins went to England in 1854 as agent of the Clinton County (Ohio) Cattle Company, he investigated the 1817 importation and reported favorably on its purity.[35]

That did a little to allay the first doubt, but what about the second? Who could say that the Shorthorn line, if pure in 1817, had not been polluted by the Longhorn line sometime during the twenty years afterward? Except where there was documentation in herd books (and not fraudulent documentation), no one could say for sure. A number of influential cattlemen, nevertheless, inferred no opprobrium from the term "Seventeens." Aware of this, L. B. Clay and Co. of Lexington exhibited some Kentucky-bred Shorthorn stock at the Chillicothe fair in 1835, one year after the Ohio Company importation.[36] The second best yearling heifer (owned by William M. Anderson) and the second best two-year-old heifer (owned by Felix Renick) at the 1840 show of the Ross County Agricultural Society were both listed as "out of Kentucky 1817 stock."[37] A cattleman in Trumbull County, Ohio, specifically asked Kentuckian Brutus Clay for 1817 stock.[38] William Renick, who was becoming sniffy about Kentucky cattle, admitted, apparently without sensing any incongruity, that some of his own cattle

[34] *Ohio Farmer* (Cleveland), January 6, February 24, 1855.
[35] Allen, *History of Shorthorn Cattle*, 167.
[36] See Chapter IV.
[37] *Scioto Gazette*, July 27, 1835; October 29, 1840.
[38] James F. King to Brutus J. Clay, January 13, 185[?], Cassius M. Clay collection.

were "half blood Durham and half blood Patton generally; though some of the Patton had a cross with the Kentucky shorthorn, of the importation of 1817, and one or two of them nearly or quite full blood of that stock—I mean the ancestors of those cattle."[39]

Meanwhile, a smear campaign was launched against the Seventeens. Possibly the motivation for the campaign came from some of the men who were investing in new importations and therefore wished to discredit the old. Now that Longhorns had decisively fallen from popularity, receptive listeners were found among farmers who feared that Longhorn characteristics might crop up in their cattle. The campaign burgeoned, moreover, on the pedigree craze of the 1850's.[40]

Although his father had experimented with Longhorns, William Renick later denounced them. This he did despite the fact that, as we have seen, there was a possibility of Longhorn blood in the Patton stock which, until 1834, formed the foundation stock of the Ohio Renicks' herds; his assumption apparently was that any Longhorn traits in his herds had been culled out by now. This left William free to shudder publicly at the spread of Longhorn blood through Kentucky since 1817. Although he never denounced the Seventeens en bloc in writing, he cast implication: "The long-horns [of 1817] being the most sightly animals, took the fancy of the people at first, and some of those having good stock of former importations, well nigh ruined them for the shambles by introducing the long-horns among them. . . . Unfortunately, in Kentucky in particular, the long-horns got a pretty general dissemination before they were entirely discarded, and a practice of somewhat indiscriminate breeding followed, producing about as undesirable a stock for the shambles as could well be imagined."[41] Did William Renick mean to imply that the only Seventeens of unquestionable purity were those on the north side of the Ohio River? Uncle Felix would never have made such an implication, but in 1848 Uncle Felix's restraint upon William was removed by

[39] W. Renick, *Memoirs*, 59.   [40] A. Sanders, *Shorthorn Cattle*, 177-78.
[41] W. Renick, *Memoirs*, 18-19.

death. William, moreover, except possibly for his early days on the Green River, had neven been on close terms with the Kentucky cattlemen.

Word reached the Ohio Valley that the eastern butchers were refusing to buy animals that looked in any way like Longhorns. Their flesh, the butchers complained, was very dark and tough, without any admixture of fat. In fact, the butchers laughed at these animals and dubbed them "Kentucky Red Horses." Although this prejudice died down enough by 1855 that the eastern butchers were willing to buy Texas Longhorns, it was enough in the 1840's to set off a Longhorn panic in the Ohio Valley: "In examining the cattle in different parts of our country . . . we discover traces of other and earlier importations, and now and then an animal whose general configuration, and particularly his horns, show that he has been derived from the long horns, a breed of cattle once very celebrated, is met with."[42]

William Renick later commented: "It is only within a few years [of 1868] that the Kentucky cattle have recovered from the odious character given them in the Eastern market by that stock. Not longer ago than ten years, the same cattle would command from twenty-five to fifty cents per hundred more if they could be passed off for Ohio rather than Kentucky stock— and this has often been resorted to by the drovers." Arriving in Brighton in 1842 with his drove, he was warned by the butcher not to call his cattle "Durhams," as: " 'The Durhams were in bad credit in that market.' I told him that I did not think there had ever been any of the improved Durhams in that market, and asked him if he called a certain twelve head, that came from New York the week before, Durhams. He said, 'yes.' Then, I said to him, it was evident that he did not know what the Durham was. We called such cattle Kentucky red horses, the worst kind of butcher's cattle."[43]

Stories like this fueled the Longhorn panic to such an extent

[42] *Monthly Genesee Farmer* (Rochester, New York), August, 1838.
[43] W. Renick, *Memoirs*, 57-58.

that cattlemen hustled to verify the lineage of their heretofore undoubted Shorthorn breeding stock. While cattle that looked like Longhorns probably were "in bad credit" in the eastern markets, it is hard to believe that the butchers rejected cattle simply because they were Kentucky stock; neither the complete Shelby correspondence from the eastern markets during the 1840's nor the incomplete Gay-Goff correspondence gives any hint that these Kentucky cattlemen encountered such rejections.[44]

At this juncture of affairs a wail went up from Tennessee. The Tennesseans were unhappy, to say the least, about the breeding cattle some Kentuckians had sold them. Kentucky's pridefully offered "thoroughbreds . . . English cattle . . . and Hereford Reds" were actually, stormed the Tennesseans, "mixed breeds of Herefords, Pattons, scrubs, with a slight shade of Durhams. . . . [They are cattle of] large frames, coarse limbs, 'raw bones,' long legs, bodies, and horns, big and indelicate heads and tails"—"abundant feeders but poor fatters!" While Kentuckians still averred that their state contained the finest cattle in the nation, one Kentucky agricultural paper admitted that "some few unworthy men professing the name of Kentuckians, have, fradulently, imposed on Tennesseans indifferent stock, under false pedigrees."[45]

The storm continued to crash upon the heads of the Kentucky cattlemen. While William Renick simply disparaged the Kentucky cattle, at least by implication, Governor Trimble of Ohio charged that Ohioans had been victimized in the same way as the Tennesseans: "The citizens of Ohio have . . . suffered . . . from . . . [the] impositions practiced . . . by a few unworthy Kentuckians . . . , who were willing to stamp by certificate, one fourth, one half, and three fourth bloods, with the character of thoroughbred."[46]

[44] Thomas H. Shelby papers, in University of Kentucky Library; Strauder Goff and James Gay papers, in possession of Ben Douglas Goff, Sr., Winchester, Kentucky.
[45] *Franklin Farmer*, March 14, 1840.
[46] *Franklin Farmer*, February 3, 1838.

To their dismay Kentuckians found that no one wanted their Longhorns, that the purity of their Shorthorns was being questioned, and that a few crooks among them might bring discredit upon their pedigrees. The first steps toward solution of the problem seemed to be to expose the crooks and to purge the Longhorns; an Indiana farm paper reported in 1846 that "a Kentucky farmer would now be very loath to let a bull of the much vaunted old Bakewell breed, with his straight back and long horns and fat all to itself overlaying the carcass, come within a ten-foot pole of his herd of cows."[47] So the Kentuckians set assiduously to the task of culling out Longhorn traits. No wonder Kentucky was the last state in the Ohio Valley to bring in Texas Longhorns for fattening.[48]

The Longhorn panic had had some justification, but in the extremes to which it went, it was unfounded and wrought much damage. The particular victims were the Seventeens. First men feared that the Seventeens might contain Longhorn blood; next men feared that their Shorthorn herds might contain Seventeen blood. The passage of years made these fears scientifically groundless, and what kept them alive was the decree of fashion, which had gone out against the Seventeens.[49] Just why fashion turned against them is uncertain. It might have been a conspiracy. Surely the Seventeen family deserved better; although some of these cattle probably were no improvement over the Gough-Miller cattle, other members of the family had carried off hundreds of showyard prizes. Alvin Sanders says: "In the hands of such men as Garrard, Clay, Warfield, Bedford, the Renicks, Trimble, Harrold, . . . a class of cattle sprang from this [Sanders] foundation that would have compared favorably with the best results attained by their English contemporaries, the Messrs. Booth and others—whose

[47] *Western Farmer and Gardener* (Indianapolis, Indiana), November 15, 1846.   [48] See Chapter VII.
[49] Nearly all the Longhorn genes had been culled out by the 1850's. Moreover, any Seventeen blood at all by 1855 constituted a minor fraction of the blood of any herd that had been bred over the years with the importations of fresher and unquestioned Shorthorn blood. L. Allen, *American Herd Book*, 66-67.

cattle—similarly descended—became 'fashionable.' In vain was this fact pointed out. . . . The fiat of fashion went out against them in later years, and whole herds of cattle carrying but a drop of the original 'Seventeen' blood were practically lost to the breed because of . . . unreasoning prejudice."[50]

About the same time Sanders had been making his importation, another Kentuckian, James Prentice of Lexington, imported two bulls, John Bull and Prince Regent. Both were certified to be of pure Shorthorn blood, and one belonged to the improved Milk breed. Neither had pedigrees. Nathaniel Hart of Woodford County and John Hart of Fayette (both in the Bluegrass) bought them from Prentice for $1,500.00. Many excellent herd-book animals, Allen says, have traced their lineage to John Bull and Prince Regent.[51]

Indirect importations were made to Kentucky during the twenties and early thirties via Colonel John H. Powell of the Philadelphia area. Between 1822 and 1831 Powell imported to his Pennsylvania farm twenty-four cows and seven bulls. The Powell cows did much to establish the reputation of Durham Shorthorns in America as a "dual-purpose" breed. One, Cleopatra, was sold by Powell to a Kentuckian, David Sutton, in 1833. Powell also had a cow that had been imported by a Baltimorean, and this cow, Virginia, became the ancestress of the Louan family of Ohio Valley Shorthorns. Charles S. Brent of Bourbon County went to the Powell farm and bought several head; on the way back a cow dropped a calf, which a cunning drove-stand keeper tried to steal by substituting a scrub calf (but Brent noticed the difference). Descendants of the Powell cattle became numerous in Kentucky during the 1830's.[52]

The half century witnessed a few exotic crosses. At the exhibition of the Agricultural Society at Captain Fowler's in Lexington, Kentucky, on September 30, 1819, "an uncommonly fine

[50] A. Sanders, *Shorthorn Cattle*, 178.
[51] Allen, *History of the Shorthorn Cattle*, 171; A. Sanders, *Shorthorn Cattle*, 179-80.
[52] A. Sanders, 184-86; Clotfelter, "The Agricultural History of Bourbon County," 3; William H. Perrin (ed.), *History of Bourbon, Scott, Harrison, and Nicholas Counties, Kentucky* (Chicago, 1882), 70.

eight months old calf, by a buffalo (bison) bull, and a cow, the property of Mr. George Thompson," was shown. "The fact was inserted in print 20 years since, that in the early settlement of Ohio, such crosses had been effected."[53] Brahmans from India reached the Ohio Valley, presumably by indirect importation; Thomas Eades of Bourbon County advertised a Brahman bull in 1852 to serve a few cows and claimed that he had received orders from the South for all the calves he had on hand for $400.00 each.[54]

In retrospect, the early 1830's may be seen as a turning point in the breeding story. Forty years had passed since the first scrubs had been brought into the Ohio Valley. The need for breed improvement had long been recognized. And a fitful start, beset by problems, had been made, utilizing the grading system. This system remained in use, especially in Kentucky; but in 1835 Alexander Waddle would question the purity of Mrs. Motte and the Seventeens controversy would be under way. The problems highlighted by this controversy—lack of records and how to weigh in importance the factors of appearance, performance, and pedigree—were dealt with after the early 1830's in considerable detail. The problem of fickle fashion, which had smitten the Seventeens, remained, however, as perplexing as ever.

[53] *American Farmer* (Baltimore), August 25, 1820.
[54] *Western Citizen* (Paris, Kentucky), July 9, 1852.

Chapter 3
The Cattle Kingdom, 1834-1860

IN THE TWO AND A HALF DECADES BEFORE THE CIVIL WAR THE old feeding regions of the Scioto and the Kentucky Bluegrass reached their peak production of fat cattle. The range cattlemen greatly increased their business, and corn fattening encroached on these ranges. Before discussing the causes and details of these developments in the various regions, we should mention the trend of production in the whole Ohio Valley.

Apparently cattle prices in general were better in the years 1831-1832 and 1835-1838 than for years afterward; unfortunately, records of crop conditions and of the supply of cattle in the early and middle thirties are fragmentary. During the agricultural depression of the 1840's production did not lag; in fact, William Renick says this was the heyday of cattle feeding on the Scioto. The drought years of 1838 and 1839 were followed by good corn years in 1843, and from 1845 to 1849. The year after each of these good corn years must surely have witnessed an unusual supply of fat cattle; the census figures for Kentucky in 1849 and the assessor's and census figures for Ohio in 1848-1849 bear this out. Thus, the sun and rain conspired to glut the already depressed markets in the 1840's[1]

One economist denies (while offering no explanation) that there was an important relationship between the size of the corn crop and the size of the livestock "pack."[2] The farmers of the time believed, however, that such a relationship existed; a Kentucky cattleman described the relationship in the Inner Bluegrass in 1855: "The quantity of corn is great and the surplus over the actual requirements large. On the contrary cattle and hogs are vary scarce owing to the scarcity of provision last year which caused all to sell last fall and winter who could find purchasers, and to the loss of many both cattle & hogs by starvation. . . . Many casual observers or individuals depending in their estimates upon the usual laws of supply and demand predicted a great decline in the price of stock since the abundance of the material to feed them would make them cheap—But the opposite is the fact and arises thus, almost every farmer has more food than he can consume and is looking for some kind of stock to eat it."[3] Thus, as in the Scioto Valley in 1831, an abundant corn crop in the Inner Bluegrass in 1855 put stockers in demand and sent their prices up. The ability to "reckon on" such relationships was the mark of a cattleman who could make money over the years.

Corn production remained heavy in the core Bluegrass counties (including Harrison) during the half century before 1860.[4] This was so despite the fact that the gently undulating topography of the Inner Bluegrass Basin was susceptible to erosion when put to corn; the Scioto Valley in Ross County, in contrast, has wide, flat bottoms between steep, wooded hills. Kentucky's big grazier-feeders, as has been noted, raised corn; and apparently cattleman Cassius M. Clay was unusual (though prophetic of modern times) in buying corn for feeding.[5]

The region extending south into Lincoln County (which for a while included what is now Boyle County), west through Shelby County, and east into Montgomery County, remained Kentucky's greatest beef area. Shelby County and the Inner Bluegrass counties of Bourbon, Clark, Madison, and Fayette each had 10,000-12,000 cattle in the 1840's and 1850's.[6] Some

---

[1] William Renick, *Memoirs, Correspondence, and Reminiscences* (Circleville, Ohio, 1880), 15; U. S. Commissioner of Patents, *Report, 1843*, pp. 55-56; *1844*, pp. 67-69; *1845*, pp. 178-81; *1847*, pp. 128-29; Ohio State Board of Agriculture, *First Annual Report . . . 1846*, pp. 22-60; *Second Report . . . 1847*, pp. 21-83; *Third Report . . . 1848*, pp. 26-96; *Fourth Report . . . 1849*, pp. 40-94, 116, 142-43, 167-69, 213-14, 222-24; *Seventh Census*, 1850, pp. 624-27, 862-64.

[2] Thomas Senior Berry, *Western Prices Before 1861* (Cambridge, Massachusetts, 1943), 245.

[3] J. Stoddard Johnston, Farm Book and Journal of events more especially bearing on Agriculture (1855-1865), in possession of the Filson Club, Louisville, Kentucky.

[4] *Sixth Census*, 1840, pp. 274-78; *Seventh Census*, 1850, pp. 626-27; *Eighth Census*, 1860, pp. 58-63.

[5] Wrote Clay, "I prefer buying some corn, and retaining the grass." *Ohio Farmer* (Cleveland), May 24, 1856.

[6] *Sixth Census*, 1840, pp. 274-78; *Seventh Census*, 1850, pp. 626-27; *Eighth Census*, 1860, pp. 58-63.

of these were owned by small farmers who had no slaves; others belonged to large landed proprietors. The cattle business —in which ten hands could tend 300 cattle and 1,500 acres— did not lend itself to slave labor. Though the Bluegrass cattlemen, most of whom also raised hemp, tended more and more to rely upon slave labor in the hemp operation, they rarely had more than forty slaves (although Brutus Clay had 150), and frequently only four or five.[7]

Grazing and feeding functions were still combined in the Kentucky Bluegrass after 1834. Most of the Bluegrass cattlemen tried to produce enough calves to fill their need for stockers, but some had to buy cattle to feed. These were obtained from the Kentucky rangelands or from Bluegrass cattlemen who had stockers to sell. Thus, Tacitus Clay of Owensboro offered his cousin Brutus of Paris 100 steers in 1839. Madison in particular seems to have had a surplus of stockers, probably because much of its land is "crawfishy" and not well suited to feeding.[8] In 1848 Joel Fox, who lived near White Hall in Madison County, offered to try to get Brutus some "three years old that wile do to feed in the spring," which meant to feed on bluegrass.[9] That part of Lincoln County around Danville (now Boyle County) supplied stockers to other parts of the Inner Bluegrass feeding region to the north.[10] Some of the stockers produced within the Inner Bluegrass were bartered or auctioned at Lexington's Cheapside Market on "court days."

---

[7] Richard Laverne Troutman, "Stock Raising in the Antebellum Bluegrass," *Register of the Kentucky Historical Society* (January, 1957), 55:17; James F. Hopkins, *A History of the Hemp Industry in Kentucky* (Lexington, 1951), 3, 19-20, 24, 27-28; Cassius M. Clay, interview with the author, Paris, Kentucky, June 1, 1956; Ben Douglas Goff, Sr., interview with the author, Winchester, Kentucky, April 11, 1957.

[8] Elizabeth Ritter Clotfelter, "The Agricultural History of Bourbon County, Kentucky, Prior to 1900" (thesis, University of Kentucky, 1953), 23; *Western Citizen* (Paris, Kentucky), August 31, October 26, 1849; Tacitus Clay to Brutus J. Clay, April 25, 1839, H. Miller to Brutus J. Clay, January 20, 1844, in possession of Cassius M. Clay, Paris, Kentucky.

[9] Joel Fox to Brutus J. Clay, January 1, 1848, Cassius M. Clay collection. Fox did not live on the Cassius Clay estate of White Hall, but that was his post-office address.

[10] *Western Citizen*, February 4, 1842; John Green to Brutus J. Clay, March 7, 1837, Cassius M. Clay collection.

The cattle-feeding industry of Bourbon County found leadership in such men as J. Bagg, Reuben Hutchcraft, George Bedford, Sam Offutt (until 1842), young Charles T. Garrard (at the family's estate Mount Lebanon, three miles north of Paris), and Brutus J. Clay (at his estate Auvergne, four miles south of Paris).[11] When Brutus' father, Green Clay, had died in 1828, Green's lands were divided between the two sons, Cassius taking the Madison County lands, where he built the White Hall mansion, and Brutus the 1,400 acres in Bourbon County. Here Brutus founded a herd of cattle in 1830, sent cattle to New York City in 1836, built a brick manor house in 1837, and by 1838 was trading with cattlemen in various parts of Kentucky. He bought and sold feeding cattle, and in 1855 sent 115 cattle, twenty of them very large, to the railroad cars. Brutus also became a bank shareholder, having bought stock in the Northern Bank of Lexington in 1859.[12]

In Clark County the Taylor family and Dr. Sam Martin, remembered for their importations, also did some feeding. Dr. Martin, who had a pond-studded farm at Pine Grove, about seven miles from the Kentucky River, reported in 1849 that "I sold my three-year-old steers last year for $40 each." "They had been corn-fed one winter." The neighborhood around Hancock's Branch, several miles northwest of Winchester, was devoted almost exclusively to the raising of Shorthorns.[13] Here the largest operation was probably that carried on by Strauder Goff on the Holmhurst estate left to him by his father Thomas; during some seasons Strauder put droves on the road for market every ten days. The "Renick boys" from Ohio were frequent

[11] *Western Citizen*, March 11, 1842; Lucien Beckner, "Kentucky's Glamorous Shorthorn Age," *Filson Club History Quarterly* (January, 1952), 26:41; William H. Townsend, *Lincoln and the Bluegrass* (Lexington, 1955), 150; *Ohio Farmer*, September 20, 1856.
[12] Cassius M. Clay interview; Brutus J. Clay Stock Book, 1830, Abner Cunningham to Brutus J. Clay, July 31, 1836, John Pratt to Brutus J. Clay, October 31, 1838, Brutus J. Clay Register of Names and Memorandum Book, 1854-1872, M. C. Johnson to Brutus J. Clay, March 8, 1859, Cassius M. Clay collection.
[13] Commissioner of Patents, *Report, 1849*, pp. 296-97; Beckner, "Kentucky's Shorthorn Age," 37-38.

overnight guests at Holmhurst; and when they came, there was so much talk that all work stopped. Just over Hancock's Branch from Holmhurst, Abram Renick, a gaunt-faced, bearded old man of unpretentious living, bred his famous "Rose of Sharon" herd.[14]

Adjoining Holmhurst on the north were the lands of "Graybeard" Sam Clay, a kinsman of the Bourbon County Clays. "Graybeard" Sam, who had cows from the Northern Kentucky Importing Company, never went to town without taking something to sell; if he had no livestock, anything else would do. In fact, this man, proprietor of 2,000 acres, would even cut greens to sell in town. During the court-day market in Winchester, he would sit all day sidesaddle on his horse, munching cold biscuits and writing checks for his purchases. When the other cattlemen adjourned to the Reeves House for drinks and supper, "Graybeard" Sam, who wouldn't smoke, drink, or chew, simply went home.[15]

No less colorful than "Old Abe" Renick and "Graybeard" Sam Clay, but in a different way, was courtly Cassius Marcellus Clay, who lived in the next county south, Madison, on the estate built by his father, Green Clay. Having come under the influence of William Lloyd Garrison while a student at Yale, Cassius was an abolitionist and as such was an embarrassment to his cousin Henry Clay, who himself opposed restoration of the slave trade in Kentucky and advocated gradual emancipation in the state. Whereas the Great Compromiser compromised with slavery to the extent that he reluctantly owned slaves, impulsive Cassius freed all his slaves. Later he refused to take any of his mother's Negroes—offering as explanation simply that the slaves "drink too much."[16] Cassius went into the cattle business at White Hall in 1833, when he bought a

[14] Goff interview, James Gay to Strauder Goff, April 28, 1848, in possession of Ben Douglas Goff, Sr., Winchester, Kentucky. After Strauder's death, "Holmhurst" passed into the hands of his only son, Ben. Like his grandfather Thomas, Ben occasionally traded in cattle, buying fat cattle and marketing them. *Ohio Farmer*, June 28, 1856. [15] Goff interview.
[16] Townsend, *Lincoln and the Bluegrass*, 79-87, 160-61; Clay interview. During the interview Mr. Clay read to me excerpts from Cassius' letters.

breeding cow from James Garrard II. Five years later, we know, he bought 84 steers. In general, Cassius emphasized grazing more than corn feeding. He girdled and burned undesirable trees until he had parklike fields similar to Shelby's.[17] His regular market was Cincinnati. But he did not make money, and in fact he fell badly into debt. Moreover, having freed his slaves, he failed to find an adequate replacement for this labor force; "it would be a great help to me," he wrote brother Brutus, "if you would come down here with your boys [slaves] and clean up my pastures."[18]

Such agricultural difficulties did not, however, deter hot-blooded blueblood "Cash" from becoming deeply embroiled in political and military activities in the 1840's. First there was the struggle to keep an abolitionist newspaper going in Lexington; then came the adventure of the Mexican War. In the latter instance Cassius went off and left his brother the job of disposing of his cattle; thus in the fall of 1846 Brutus tried unsuccessfully to get Walter Chenault, a big operator who sent cattle occasionally to market at Charleston, South Carolina, to take Cassius' cattle. Upon his return from Mexico, instead of quietly resuming agricultural pursuits, "Cash" plunged again into politics, this time as candidate for Madison County delegate to the constitutional convention of 1849; Cassius was standing as candidate of the emancipationist wing of the Whig party, that is, the group which proposed constitutional machinery for eventual emancipation. Cattleman Edwin Dudley of Bourbon was standing for the same cause, but he never got into such trouble as candidate Cassius, who, although adept at defending himself with a bowie knife, was beaten up and stabbed to within an inch of his life.[19] Finding that he had to

[17] Brutus J. Clay Stock Book; Cassius M. Clay to Brutus J. Clay, March 21, 1838, Cassius M. Clay collection; *Ohio Farmer*, May 24, 1856.

[18] David Smiley, "The Public Career of Cassius M. Clay" (dissertation, University of Wisconsin, 1953), 52; Clay interview.

[19] Townsend, *Lincoln and the Bluegrass*, 99-119, 157-75; James Gay to Strauder Goff, April 28, 1848, B. D. Goff collection; Walter Chenault to Brutus J. Clay, October 23, 1846, Cassius M. Clay collection; *Western Citizen*, May 11, 1849.

run Cassius' farm most of the time anyway, Brutus eventually made a partnership with him. This partnership remained in effect while Cassius was ambassador to Russia (1861-1867) and was not terminated until 1869.[20]

In Fayette County large feeding operations were carried on by the Dudleys, Jacob Hughes, Tom Shelby, Nat Hart, John Kerr, Otho Offutt, and the partnership of Edward S. Washington, A. D. Offutt, and Glass Marshall. Kerr's extraheavy cattle were exhibited before presumably admiring butchers in Cincinnati. Otho Offutt produced mules, saddle horses, and fat cattle on his plantation in the Elkhorn Valley before his death in 1841; many of the cattle and mules were driven to Natchez or taken on steamboats to New Orleans.[21]

Hughes, whose manor house Leafland may be seen today on the Lexington-Winchester road, owned the Great Northern Bank in Lexington, sat in the legislature, and marketed cattle in the East and hogs in Cincinnati. He did not as a rule raise his own stock cattle; "I buy," he said, "graze and feed about 300 cattle annually; raise and sell about 200 hogs." On his farm of 1,800-1,900 acres he had 200 acres of corn, 20 acres of meadow, 100 acres of wheat and rye, and the balance in pasture. He worked ten hands. His profits, he reported, were $9,945.00 in 1835 and $10,475.00 in 1836. Later, however, he went broke, but recouped his fortune by buying two ferryboats at Covington and charging hog drovers $1.00 per 100. Hughes is said to have dealt also in slaves (perhaps acquired in foreclosures to his bank), keeping them behind bars in the cellar of Leafland.[22]

Over in Franklin County, the most famous cattleman was Robert W. Scott. City-bred Scott took over a worn-out, badly arranged farm and restored it by means of manuring, rotations,

[20] "Agreement to Terminate Cattle Partnership of Cassius M. Clay and Brutus J. Clay, 1869," Cassius M. Clay collection.
[21] *Franklin Farmer* (Frankfort, Kentucky), September 23, December 16, 1837; *Western Citizen*, March 4, 1842; *Scioto Gazette* (Chillicothe, Ohio), September 24, 1840; Townsend, *Lincoln and the Bluegrass*, 150-51.
[22] *Franklin Farmer*, September 23, 1837; Goff interview.

and experimentation. Hemp was his chief cultivated crop.[23] In weedless pastures and behind numbered gates, he fattened fine Durhams. Though his six-year-old cow Lady Gray weighed 1,500 pounds, and his two-year-old bull Frederic 1,680, he called none of these really fat. He usually had several cattle belonging to his neighbors on pasture at his farm; the fee for this accommodation in the late 1830's and early 1840's was $2.00 per head per month, except for calves, which were $1.50. On April 4, 1839, H. Blanton put thirty-nine head of cattle and eight horses "to keep" at Scott's on straw and fodder, an indication either that the drought of the previous summer had seriously damaged the bluegrass pastures or else that the cattlemen were conserving them (which would have been wise, as another drought was approaching). Blanton agreed to pay a fee of $1.50 per head "to be paid in plank." By the late 1850's, although Scott still had, he said, "a lot of fine young bulls and heifers," he was buying beef, apparently having reduced his own cattle operations.[24]

The competition which the Kentuckians found Ohio offering in the eastern cattle markets in the 1830's was undoubted evidence of the growth of the cattle-fattening industry in the Scioto Valley. In 1835 a regular stock cattle market opened in Chillicothe, but some stockers were also probably handled by merchants like Joline and Co., on Main and Mulberry at the canal. The practice of selling corn to cattle feeders continued; in one instance the use of "two sets of convenient feeding lots" was included with the price of the corn, and experienced men to do the feeding could "be hired on the premises." In this instance the man who was selling the corn had not raised it but admitted he had bought it as a speculation.[25] The value of cattle sent east by Scioto Valley feeders in 1838 was put at

[23] Hopkins, *Hemp Industry*, 23; Herbert A. Kellar (ed.), *Solon Robinson, Pioneer and Agriculturist* (Indianapolis, 1936), 1:249.

[24] *Kentucky Farmer* (Frankfort), July, 1858; *Franklin Farmer*, October 28, 1837; Robert W. Scott Farm Daybook, 1834-1859, Filson Club.

[25] Charles T. Leavitt, "The Meat and Dairy Livestock Industry, 1819-1860" (dissertation, University of Chicago, 1931), 94; *Scioto Gazette*, September 9, 1835; April 9, 1840.

$700,000.[26] That year, however, there was a severe drought, which reduced the corn harvest in the fall, and thereby reduced the export of fat cattle in 1839, when "in the Sciota Valley in Ohio, where from ten to fifteen thousand head, are usually fed for market, the farmers were unable to furnish more than three or four thousand in the whole."[27] A second drought year followed, and to add to the troubles of the Scioto Valley cattlemen, depression was reaching west into the Ohio Valley. The expensive breeding stock that many of them had on their farms was now declining in value. The Whigs tried to make political hay by warning the Scioto feeders in 1843 that if the Whig tariff were repealed, they might be pinched between high prices for stock cattle and low prices in the eastern markets for fat cattle.[28] Meanwhile, production increased, and cattle feeding on the Scioto reached its height during the 1840's.[29]

Although 1845 was a good corn year in the Scioto Valley, a severe drought struck the Western Reserve, and the dairy farmers there were wondering how to get their cattle through the winter. The *Ohio Cultivator* offered a suggestion, which a few of them accepted: winter your cattle in the west-central Ohio upland or in the Scioto Valley. The paper reported that Scioto corn standing on the ground, warranted to yield an average of fifty bushels of shelled corn per acre, could be purchased for $9.00 to $10.00 per acre, with half a dollar an acre extra for cutting and putting into shocks. Fields of lighter quality, yielding forty bushels per acre, could be had for $7.00 to $8.00 per acre. A cheaper way to winter cattle in the Scioto Valley was to buy only the corn fodder after the corn was husked; this could be contracted for in August for about 7 cents a shock of twelve hills square, or for $1.37 to $1.50 per acre for that having a full growth of stalks. The Scioto Valley

[26] Leavitt, "The Livestock Industry," 100.
[27] *Monthly Genesee Farmer* (Rochester, New York), July, 1839.
[28] *Scioto Gazette*, June 22, 1843. Actually, this was a specious argument, for the tariff of 1842 contained no schedule on cattle or beef. 5 *U. S. Stat.* 548-67. Thus, when a temporary improvement in cattle prices did occur in 1843 (see Chapter VI), the tariff was not responsible.
[29] W. Renick, *Memoirs*, 15.

farmers were reportedly disinclined to care for Western Reserve cattle at a set price per head for the season. This meant that the Western Reserve farmer would have to spend the winter with his cattle in the Scioto Valley.[30]

By 1846 George Renick, now seventy years old, decided to retire from the cattle business. On July 29 he held an auction on his Paint Hill estate overlooking Chillicothe and sold all his cattle except five which he reserved for himself. The number of bidders was "respectable, though not as great as we expected to have seen; nor were the prices at which the cattle sold generally as high as we supposed stock of such pedigrees would bring; though much allowance must be made for the times. We believe Mr. Renick is not at all disappointed with the result of the sale."[31]

Two years later, George's brother Felix died in an accident at a ferryboat dock. But leadership of the cattle industry in Ross County was still provided by men like Dr. Arthur Watts, who owned 1,000 acres, and Edwin Harness, who had gone with Felix to England in 1833 and whose estate High Banks was just across the Scioto River from Felix's Indian Creek farm. Ross County in the late 1840's had some 15,000 beef cattle excluding milch cows and working oxen. In Pickaway, the next county north of Ross, cattle fattening was carried on by George Renick's sons, William (of Mount Oval overlooking the Pickaway Plains) and Harness, by J. P. Brown, and by some German farmers along Deer Creek—John Walke, William Miller, and Henry Baker. Pickaway in the late 1840's had probably 19,000 beef cattle.[32]

James Inskeep Vause and Jacob Vanmeter, both of whom had bottom land along the Scioto, were probably the only important cattlemen in Pike, the next county south of Ross. Jacob Vanmeter advertised 500 fat cattle to be sold October 1-15,

---

[30] *Ohio Cultivator* (Columbus), August 15, September 1, 1845; April 15, 1846. [31] *Ohio Cultivator*, August 1, 1846.

[32] *Ohio Farmer*, August 2, 1846; Ohio Board of Agriculture, *First Report . . . 1846*, pp. 59-62; *Sixth Report . . . 1852*, pp. 51-53; *Ohio Cultivator*, May 15, 1849; Ohio Assessor's census of farm animals for 1848, in Commissioner of Patents, *Report, 1848*, p. 22; *Seventh Census*, 1850, pp. 862-64.

1848. On September 17, 1851, Vause paid $1,077.45 for what we can deduce must have been a herd of about 100 stockers to put to corn that winter. He and the Harness family (who were his in-laws) traded back and forth with Henry Baker of Pickaway County. Vause's relative, A. A. Inskeep, raised cattle at Moorehead, on the South Branch of the Potomac, and had pastureland "in Allegheny";[33] hence Vause may have continued the historic Scioto–South Branch cattle trade. During the turbulent days of the breakup of the Whig party and the approach of the Civil War, Vause, who by this time had acquired a residence in Jefferson Township, Ross County, followed Henry Clay's son James into the Democratic party and had a sharp falling-out with fellow cattleman William Renick of Mount Oval, who became a prominent propagandist of the Republican party.[34]

The middle and late 1850's found the Scioto Valley feeders in a vulnerable position. Demand from distilleries and from abroad ran the price of corn up to about forty cents a bushel. Purchase of corn for feeding was unthinkable, and cattlemen who had corn found it much more profitable to sell than to feed it—an exact reversal of the situation of thirty years before. The railroads, furthermore, were enabling Illinois feeders to send cattle to New York at little more cost than the Scioto feeders could; yet prices in the New York market from 1856 through 1858 were not considered good. The Scioto feeders, therefore, turned to a new system, "half feeding." This meant that the cattle were fed on corn only the latter half of the winter, having been kept on hay and other fodder until then. In March they were put out to grass and then were shipped to reach market from the first of June to the first of October.[35]

By 1860 the number of beef cattle in Ross and Pickaway

[33] *Scioto Gazette*, September 27, 1848; J. I. Vause Account Book, 1849-1876, A. A. Inskeep to James Inskeep, May 12, 1856, in possession of Joseph Vanmeter, Chillicothe, Ohio.

[34] Joseph Vanmeter, interview with the author, Chillicothe, Ohio, February 29, 1956.

[35] Robert Leslie Jones, "The Beef Cattle Industry in Ohio prior to the Civil War," *Ohio Historical Quarterly* (April, July, 1955), 64:317-19.

counties had been reduced to about 14,000 in each county, an average reduction of about 3,000 head per county from the 1848-1850 levels. Except for this not too drastic reduction in volume and the new practice of half feeding, the beef-cattle industry in the Scioto Valley remained very similar in 1860 to what it had been in the 1840's. The feeders still made trips to the Kentucky Bluegrass to select stockers, as Felix Renick had done in 1816. However profitable cash grains may have been, and despite the wheat upsurge in southern Ohio in the 1850's, the Scioto Valley did not become a cash-grain region; in fact, it imported cash grains from the Kentucky Bluegrass.[36]

The Miami Valley, meanwhile, developed after 1830 into an important cattle-fattening region which, though offering only occasional competition to the Scioto feeders in the Pittsburgh and eastern markets, met the Kentucky cattlemen head on in the Cincinnati market, to which the Miami Valley was a natural hinterland. A pattern developed in which the Miami Valley brought in stockers for feeding, exported yearling steers to the west-central Ohio graziers, and exported cow calves to the Western Reserve dairymen.[37]

Warren County for the first time in its history began to have a concentration of capital in the 1850's. Here the Hollingsworths and Steddoms had corn and livestock farms totaling 1,600 acres. Moses Steddom butchered at least some of his own beef, selling some, for example, in November and December, 1853, to neighboring farmer Thomas Longstrith on credit; Moses had a constant stream of hired hands splitting rails, covering corn, etc. Jacob Egbert had at least several thousand dollars invested in cattle and hogs in 1852.[38] The Union Village Shakers owned 4,000 acres, including excellent woodland pas-

[36] *Eighth Census*, 1860, pp. 112-17; Samuel Adams to Brutus J. Clay, April 2, 1858, Cassius M. Clay collection.

[37] Josiah Morrow Scrapbook, in possession of the Philosophical and Historical Society of Ohio, Cincinnati; Jones, "The Beef Cattle Industry," 291.

[38] Tax Duplicate, Warren County, 1851, 1852, 1855, in Auditor's Record Room, Warren County Courthouse, Lebanon, Ohio; L. H. Everts, *Combination Atlas Map of Warren County,* Ohio (Philadelphia, 1875), 65; Moses Steddom Account Book, in possession of the Warren County Historical Society, Lebanon.

THE CATTLE KINGDOM, 1834-1860 53

ture. The society was divided into four principal families, each having a herdsman and an agent. They were said in 1850 to have 280 "high-grade short-horns," and in 1856, 450-500 head of blooded cattle.[39] Warren County had in the late 1840's normally 7,000-12,000 cattle, of which 5,000 were beef cattle intended for export to graziers or butchers.[40]

Clermont County, with its varied soils and topography, was a mixture of rangeland and feeding areas. The hilltop and plateau farms tended to be retarded; although there had been some gradual breed improvement, a farmer at Lindale (on the hills north of New Richmond) remarked as late as 1870 that "we have only common cows."[41] On the other hand, cattle fattening was conducted by the Salts, Hitches, Wests, Elys, and Shotwells along the East Fork of the Little Miami. The county exported in the 1840's about 1,000 head of fat and stock cattle annually, valued at $20,000. Fertile Butler County had about 11,000-12,000 head of cattle.[42]

In the upper Miami Valley, Montgomery County (around Dayton) reported in the late 1840's nearly 14,000 head of cattle, and Steele, Inskeep, Partridge, and Judge Holt were mentioned as cattle feeders. Greene County had in 1849 about 10,000 cattle. Western Clark County, lying in the Miami Valley around Springfield, engaged in fattening, and sent fat cattle to Cincinnati and Pittsburgh. Miami County, far enough north to be bothered by the drainage problem, did but little cattle fattening.[43]

Slightly less important than the Miami Valley as a cattle-feeding region was the Wabash Valley belt of corn farms. By 1840 this belt had twice as many farm workers per square mile

[39] *Ohio Farmer*, June 7, 1856; Morrow Scrapbook.
[40] Ohio Assessor's census, 1848, 1849; Morrow Scrapbook.
[41] William Latman to Brutus J. Clay, February 28, 1854, J. S. Galloway to Brutus J. Clay, April 14, 1870, Cassius M. Clay collection.
[42] Ohio Board, *Third Report* . . . *1848*, pp. 26-28; *Fourth Report* . . . *1849*, pp. 66-67, 222-24; *Fifth Report* . . . *1850*, p. 295.
[43] Ohio Assessor's census, 1848, 1849; *Ohio Cultivator*, July 15, 1845; Ohio Board, *Third Report* . . . *1848*, pp. 26-28; *Fourth Report* . . . *1849*, pp. 170-73, 222-24; Oliver S. Kelly, "Springfield as Remembered Sixty Years Ago," *Yester Year in Clark County* (Springfield, Ohio, 1947), 1:30; Morrow Scrapbook.

as the adjoining country to the west, north, and east, and corn was the major crop, accounting for more than 70 percent of its production. In 1850 Putnam County recorded 9,432 cattle other than milch cows and working oxen; Hendricks, 8,439; Montgomery, 11,246; and Tippecanoe (part of which lay within the Grand Prairie), 12,851.[44]

The raising of stock cattle and working oxen for sale was based chiefly on grazing the native praries, some of which had been fenced, and perhaps bluegrass fields in winter. Improved grasses, such as timothy, herd grass, and orchard grasses, were receiving some attention. Corn was grown both on bottom lands and on tablelands. The farmers commonly put the same field to corn as many as fifteen years without intermission and without manure. The average estimated yield of sixty-five bushels per acre in bottom land is probably too high to be accepted as average for the whole region. Fat cattle, then fat hogs, then stock cattle and stock hogs, were admitted to the cornfield in that order. Both cattle and hogs would consume much of the stalk. The feedlot or stall feeding of hay and fodder was also of some importance.[45]

A fifth feeding region, the Sangamon Valley of Illinois, exhibited actually a mixture of stall feeding and open-range grazing. Because this region into the 1830's had been part of the Illinois range and continued into the 1850's to be partly range, it will be included in the discussion of ranges.

The literature of the time gives bits of evidence that the corn-and-livestock economy was beginning to exhaust the soil in some places in the Ohio Valley and that a few people were aware of what was happening. Some Ohio farmers regarded manure as a nuisance, and the more uningenious of them could think of no way to dispose of it except to haul it in a wagon to the public road and dump it there. John Johnson wrote, "I am much mistaken if a great deal of the land in Ohio does

---

[44] Arlin D. Fentem and Robert Reece, "A Preliminary Study of Agriculture in the Old Northwest in 1840," ms. in possession of the authors, University of Wisconsin; *Seventh Census*, 1850, pp. 790-92.

[45] Fentem and Reece, "A Preliminary Study of Agriculture."

not now [1858] need manure. . . . I did not see a manure heap in Ohio, only where cattle had stood or been fed in the woods, and the manure left there to waste." William Henry Harrison claimed that if proper rotations of crops were practiced, manuring was unnecessary.[46] Two years in grass rested the land from corn; and if cattle were grazed on it, they automatically manured the land.

The fact that the soil of the Scioto Valley has been well preserved suggests that the better cattlemen here must have practiced manuring; corn yields of 100 bushels per acre with "manure and good attention" were reported in the early 1840's.[47] At least one Kentucky cattleman, Robert Scott, manured his fields, and with good effect. Wrote central-Kentuckian Sam Martin, "Those who scrape up their manure . . . do not live about here."[48]

The ranges, six in number, were the Kentucky range, the southern-Ohio range (on both sides of the Scioto Valley), the west-central Ohio upland, the artificial grasslands of Darke and Mercer counties, the Grand Prairie of Indiana northwest of Lafayette, and the Illinois range. By 1860 feeding had advanced into four of these.

Selecting the stockers required skill on the part of the feeder. He could tell the age of the cattle by the teeth or by the horns. There are no teeth until the animal nears its second birthday, when two teeth appear; two more teeth are cut at three years of age; two more at four; and two more at five, when the animal is called "full-mouthed," though the last two teeth are not fully up until six years of age. At three years of age the horns are smooth and handsome; then comes a circle or wrinkle, and a new circle is added every year. Therefore, the first circle represents the fourth year of age. A range-cattle dealer seeking to deceive a feeder might file off these circles. Not to be con-

---

[46] William O'Bryan, *A Narrative of Travels in the United States of America* (London, 1836), 133-34; *Kentucky Farmer,* October, 1858; *Franklin Farmer,* November 25, 1837.

[47] Commissioner of Patents, *Report, 1845,* pp. 179-80.

[48] Hopkins, *Hemp Industry,* 23; *Maysville Eagle* (Maysville, Kentucky), May 14, 1845.

fused with these circles are the ringlets found at the root of the horn—an indication that the animal had been ill-fed during its growing period. Cattle were bought as stockers at any age from one to four. In that era when many cattle feeders desired to produce extraheavy beef, the stockers themselves might average 800 pounds, as did the forty-seven that Brutus Clay bought for $24.00 per head in 1854.[49]

Eventually a stock-cattle market arose in Paris when such cattle from the Kentucky range to the east and north were brought into town for sale in the courthouse square during the sessions of the Bourbon County Court.[50] Thin cattle are noted as plentiful in Kentucky in 1841, though still high-priced. In 1856, which was very nearly the peak of the beef boom, stockers brought $12.00 to $34.00 per head at Bourbon County court day on March 4. Shorthorn bulls which the Bluegrass feeders had culled out for one reason or another were available to rangeland farmers.[51]

Far to the west was another part of the Kentucky range, the lands beyond the Tennessee River, which had belonged to the Chickasaw nation until 1818, and where in the 1820's young Cassius and Brutus Clay had hunted on their father's absentee landholdings. Now the newly erected Marshall County raised a "scrub breed of cattle, [which] having more industry, do better on our wild grass." "Unmarked" cattle were still wandering along the Mississippi River in 1856-1857; the red coloring of many of these suggests Shorthorn blood.[52]

Feeding made a hesitant advance into the Kentucky range.

[49] *Farmers' School Book,* quoted in *Franklin Farmer,* October 14, 1837; *Ohio Farmer,* May 31, 1856; Brutus J. Clay Register.

[50] *Kentucky Farmer,* December, 1858. Court day was the first Monday of the month. Sales were executed from 8 A.M. to 1 P.M., at which hour everyone adjourned to the Duncan Tavern, also on the square, for lunch. As much as $250,000 worth of cattle, horses, and mules would change hands, on credit (bankable promissory notes). Rudolph A. Clemen, *The American Livestock and Meat Industry* (New York, 1923), 75-77.

[51] William P. Hart to Virginia Shelby, August 27, 1841, in possession of the Filson Club, Louisville, Kentucky; *Ohio Cultivator,* March 15, 1856; *Franklin Farmer,* November 9, 1839.

[52] Clay interview; *Valley Farmer* (St. Louis), January, 1857; Johnston Farm Book.

By 1840 Warren and Barren counties (in the "barrens" between the Green and Cumberland rivers) were doing some cattle feeding; these two counties by 1850 were producing as much corn as some of the Inner Bluegrass counties had in 1840. But some local people in the Green River barrens still stated that grass could be the only basis for their cattle industry: "Our section [south of Green River] can never be a stock purchasing or stock raising country to any extent until we change our system of farming we must grass our lands and plough less." Marshall County, meanwhile, was no longer entirely range but by 1849 was shipping some fat beef "on foot" to New Orleans.[53]

The hill country of southern Ohio continued to furnish some of the stockers for the feedlots of the Scioto Valley. Athens County exported 5,000-6,000 stockers in 1849. Feeding never made much of an advance into the hill counties of southeastern Ohio; these counties had only one-third to one-half the agricultural production of the counties immediately to the north, and agriculture, although intensive, was severely handicapped by topography. Some cattle feeding did occur west of the Scioto, in Highland County, where former Governor Trimble was active in the cattle industry; this county had some 10,000 cattle in the late 1840's. But Adams, Brown, and southern Highland counties remained largely a stocker country.[54]

The 1830's and 1840's saw the graziers of the west-central Ohio upland rise to prominence. A Mr. Gwinn in 1832 had a farm of 4,000-5,000 acres enclosed in large fields for the support of 1,200 head of cattle. In 1838 David Selsor went into the grazing business at the headwaters of Paint Creek. Colonel Peter Buffenberger had a grazing farm on the "buck prairie" just over the Madison County line from South Charleston: "a perfect speciment of a purely grazing or cattle farm. It embraces 2000 acres of beautiful smooth prairie land . . . with a few large trees. . . . Of this land he has twelve hundred acres

[53] *Sixth Census*, 1840, pp. 274-78; *Seventh Census*, 1850, pp. 624-27; Commissioner of Patents, *Report, 1849*, pp. 296-97.
[54] Ohio Board, *Fourth Report . . . 1849*, p. 43; Fentem and Reece, "A Preliminary Survey of Agriculture."

in one field. . . . In this field he has 400 head of cattle, mostly full grown, and many of them one-half to three-fourth blood Durham, purchased when one or two years old from farmers in surrounding country, and here kept until ready for the butchers or drovers." Buffenberger, George Linson, and Selsor owned a total of about 10,000 acres.[55] The ratio of prairie acres to cattle was evidently about three or four to one.[56]

The grazier's operation often included mules, horses, sheep, and goats; one grazier had 400 ewes and 25 bucks. These graziers of the west-central Ohio upland obtained their cattle from outside sources, some from Warren County (in the Miami Valley), more from Illinois and Missouri; the National Road brought these latter cattle to the heart of the upland. The cattle were purchased for grazing at two or three years of age and occasionally as yearlings. A buck-prairie grazier such as Gwinn or Buffenberger would snort that the production of calves "don't pertain to my business . . . that is done by small farmers."[57] Some graziers, however, did keep cows to produce calves. Ross County cattle were still being summer-pastured in the upland; three pastures totaling 1,500 acres on the North Fork of Paint Creek, Fayette County, were offered for rent in 1835, two pastures for the season, and one by the month at one shilling per head of cattle. During the 1840's cattle belonging to farmers outside the upland were occasionally wintered on grass and hay in Madison County for forty cents a month.[58]

Madison and Fayette county graziers about 1850 were pasturing at least 25,000 head of cattle per year from the West (Indiana, Illinois, Missouri, Iowa, and Wisconsin). About 20,000 head of cattle, three years old and upward, were sold

[55] Leavitt, "The Livestock Industry," 92; David Selsor Graham, interview with the author, Midway, Ohio, September 9, 1955; *Ohio Cultivator*, August 1, 1845; *History of Madison County, Ohio* (Chicago, 1883), 831.

[56] This contrasts with a minimum of five acres per steer in the shortgrass country of Dakota and a minimum of fifteen in the shortgrass country of Texas. J. Russell Smith and M. Ogden Phillips, *North America* (New York, 1942), 507.

[57] *Scioto Gazette*, October 13, 1836; Morrow Scrapbook; Jones, "The Beef Cattle Industry," 180.

[58] *Scioto Gazette and Independent Whig*, April 15, 1835; *Ohio Cultivator*, August 15, 1845.

and driven from Madison County during seven months of 1848. Ordinarily, Madison County cattle went to the Cincinnati market, some to the Scioto feeders, some to the eastern market cities, and more to the feeders of New York and Pennsylvania. Pasturing was still the chief livelihood of Fayette County in 1848. Clark County, the eastern half of which belongs to the upland and the western half of which belongs to the Miami Valley, exported 3,000 head of cattle in 1850, "mostly sold to feeders on Scioto and driven east."[59] A regular stock-cattle market was opened in the heart of the range, at London, seat of Madison County, in 1855.[60]

The graziers of the west-central Ohio upland carried on a similar operation to that of the cattlemen of the Great Plains in a later day, and they did so in a somewhat similar environment. There the sparsely settled plains were but gradually being fenced. No Ohio statute required cattle to be fenced in. A man whose crops were ruined by roaming cattle had no recourse unless the "fence viewers" were convinced that he had erected a reasonably sufficient fence to keep cattle out. So many cattle were still roaming about unfenced in the upland in the 1850's that trains on the Little Miami Railroad between Xenia and Columbus were reported to have killed 1,600 head within four years.[61]

There were even "cow towns"—London, Springfield (during the brief period when it was the western terminus of the National Road), and especially South Charleston. Here drovers, having put up at the Ackley Hotel, assembled herds of range cattle and fat cattle and took them onto the nearby National Road. In and around South Charleston lived two cattle dealers

[59] Jones, "The Beef Cattle Industry," 184-85; Ohio Board, *Third Report . . . 1848*, p. 48; *Fifth Report . . . 1850*, p. 284.

[60] Leavitt, "The Livestock Industry," 180. Cattle sales at London became a kind of fair, where farmers talked politics and examined the wares of implement dealers. Jones, "The Beef Cattle Industry," 301.

[61] *Ohio Farmer*, January 9, 1858; Clarence H. Danhof, "The Fencing Problem in the Eighteen-Fifties," *Agricultural History* (October, 1944), 18:170. Lawsuits arising from such slaughter reached the Ohio supreme court, which ruled in 1856 that if the conductors were negligent, the railroad company was liable for damages. *Ohio Cultivator*, October 1, 1856.

and twenty-seven cattle raisers. No cattle king of the Far West ever made a more colorful entry into town than did Colonel Darlington Pierce, who for years after the Civil War would wear a broad-brimmed hat and ride down the main street of South Charleston on a white horse.[62]

The graziers of Madison County, noted the shrewd Aaron Miller in 1832, "make money fast at the business, from what I can learn they make about fifty percent clear on their money annually." This estimate, based perhaps on local tavern gossip, was probably somewhat high. To estimate the profits at 25 to 33 percent would probably be closer to the mark.[63]

Grass grazing, meanwhile, was slowly losing ground to corn feeding in the upland. Possessing land along Paint Creek in Fayette County, General Beatal Harrison probably corn-fattened some of the cattle in which he dealt and which he drove east over the mountains. The native forest of Fayette County had been almost entirely cleared by 1840. Wheat was relatively more important in Fayette than in the Miami Valley, and stock cattle were sometimes wintered there on nothing but a straw pile.[64] By 1854, apparently enough corn was being grown along Darby and Deer creeks in the upland to support some feeding there. David Selsor, of southern Madison County, bought corn from neighbors to supplement his grazing operations. Selsor's fancy corn-fed "Christmas beef" won several trophies in New York City.[65] A few miles to the northwest of Selsor's outfit was the village of South Charleston, where drovers assembled herds of fat cattle. As early as 1845, William Pierce and the Harrolds were doing cattle feeding just south of South Charleston, and the vicinity of this town became also a center of breed improvement in the upland. Western Clark County had always done

[62] *Ohio Cultivator*, August 1, 1845; Graham interview; L. H. Everts, *Illustrated Historical Atlas of Clark County* (Philadelphia, 1875), 93; J. L. Carr, interview with the author, South Charleston, Ohio, September 9, 1955.

[63] Jones, "The Beef Cattle Industry," 183.

[64] L. H. Everts, *Illustrated Historical Atlas of Fayette County* (Philadelphia, 1875), 7; Fentem and Reece, "A Preliminary Survey of Agriculture"; Jones, "The Beef Cattle Industry," 289-90.

[65] *Ohio Railroad Guide* (Columbus, 1854), 43-44; Graham interview.

cattle feeding; now that activity had spread eastward to the South Charleston area, where the county lines of Clark, Madison, and Fayette counties come together. Yet in 1850 there were still some 5,000 or 6,000 western cattle on open range in Clark County, and a Mr. Ryan of Jefferson, in northern Fayette County, as late as 1858 was grazing two-year-old Bourbon County (Kentucky) Durhams on 700 acres of "choice stock land." Over much of the west-central Ohio upland, corn growing and cattle fattening awaited solution of the drainage problem, which was being met in the upland by such means as "gopher ditches."[66]

In 1834 the development of the Grand Prairie of Indiana, still a largely unsettled range, was about to begin. Henry L. Ellsworth, commissioner of patents in the 1830's, arguing through pamphlets, farm journals, and a wide correspondence, was encouraging this development. He bought his first land in Indiana in 1835; in 1845 he made his home in Lafayette and soon was experimenting with farm machinery, fencing materials, crop rotation, and methods of livestock feeding. In 1849 he was reportedly getting an average of sixty bushels per acre from his 1,200 acres of corn land. He exported cattle via the Wabash canal and brought lumber back from Toledo.[67]

Some of the farmers on the prairie sowed a "sod crop" the first year. Five good yoke of cattle with a large breaking plow, said Ellsworth, would break up one and a half to two and a quarter acres per day, according to the weather. Cattle were wintered on bluegrass (if it lasted into the winter) and corn; the corn in their stomachs and the shelter provided by corn stalks in the field were thought to keep the cattle warm enough that they did not need sheds. In areas where little corn was raised, the cattle were put in sheds for the winter, where they were fed stacked prairie hay with perhaps a small ration of corn

[66] *Ohio Cultivator*, August 1, 1845; September 1, 1858; Jones, "The Beef Cattle Industry," 181; Graham interview.

[67] Sherman N. Geary, "The Cattle Industry of Benton County," *Indiana Magazine of History* (March, 1925), 21:28; Paul W. Gates, "Hoosier Cattle Kings," *Indiana Magazine of History* (March, 1948), 44:1, 4-6.

or meal or corn fodder. From May to December herdsmen kept the cattle on the open-prairie range, where all that had to be provided for them was salt.[68] Moses Fowler (who came in 1839), Edward Sumner (who came in 1846), and Adams Earl did both grazing and feeding on their huge properties (up to 30,000 acres per man). Fowler owned a whole fleet of steamboats plying between Lafayette and New Orleans and fattened as many as 2,000 cattle at a time. Earl (for whom the town of Earl Park on U. S. 52 is named) had a distillery, the slop from which he fed to livestock. Sumner had so many cattle (about 2,000) and used so much feed, that neighboring farmers raised corn specifically to supply this need. By 1850, however, most of the Grand Prairie remained range; Tippecanoe County had 12,851 cattle, but Benton, the heart of the prairie, only 2,379.[69]

The open-prairie range of Illinois, likewise, was slow to be broken to corn farming. Settlers preferred the "dry" prairie over the "wet," but very little even of the "dry" prairie had been taken up by 1840. One deterrent may have been the relatively high cost of trained oxen ($50.00 to $80.00 per yoke) or the $2.00 to $3.00 per acre cost of hiring someone to break the prairie. It cost an estimated $15.00 to dig a thirty-foot well. Settlement followed the timber-fringed streams, locating on the terraces rather than the low and often wet and unhealthful flood plains. Since there was no state statute requiring livestock to be fenced, many farmers were afraid to break the prairie, because they would either have to haul wood a long distance for fencing or else run the risk of having their unfenced crops trampled by range cattle. Wood fencing probably cost more than $700.00 per mile in Illinois.[70]

Moreover, the open-range methods were still profitable in the early 1850's. Even if the prairie pasture did not suffice and

[68] Commissioner of Patents, *Report, 1843*, pp. 120-21; *1845*, pp. 383-85; *Renssalaer Gazette* (Renssalaer, Indiana), May 25, 1859.
[69] Gates, "Cattle Kings," 8, 13-15; *Seventh Census*, 1850, pp. 790-92.
[70] Fentem and Reece, "A Preliminary Survey of Agriculture"; Danhof, "Fencing Problem," 171-72.

it was necessary to buy winter fodder in the form of timothy, clover, or prairie hay, the usual fee was only $2.00 to $2.50 per head. Illinois prairie-fed and foddered cattle of three to five years of age were ready for the market at a cost of $6.00 to $12.00 per head for the producer. They could be sold for up to $25.00 per head.[71] The Illinois range, open and unfenced, thus survived into the 1850's, even within sight of Springfield, in the Sangamon Valley. The groves were said to be infested with robbers and counterfeiters; and in 1855 an incident of cattle-rustling occurred.[72]

Corn feeding, however, was encroaching on the range, particularly in the Sangamon Valley, but also in McLean County, which was the site of Isaac Funk's cattle kingdom and which in 1850 had 15,000 cattle and three times as much corn as Champaign County. The corn farmers, if they did not plant a "sod crop" the first year, broke the prairie by plowing in June or July and let the sward lie until the next spring. They planted the corn in hills. When the corn plants were three to four inches high, they began to plow the field and continued until the corn began to "set for" ears; then it was "laid by," which meant that nothing more was done except to chop down large weeds. Many of the "sloughs" which were avoided by the first settlers became "dry" prairies merely by exposure to sun and air through annual mowing and grazing of the tall "slough" grass.[73]

Technological advance helped too. As trainloads of lumber rolled south from Chicago, fairly comfortable farmhouses could be built on the prairie, far from any woods, for $200.00 to $1,000.00. The cobs from a hundred acres of corn would supply a family's fuel needs for a year. Pine logs for worm fences

---

[71] Theodore L. Carlson, *The Illinois Military Tract, A Study of Land Occupation, Utilization, and Tenure* (Urbana, 1951), 129. Although the Military Tract is not in the Ohio Valley, prairie conditions there obtained elsewhere in Illinois too. Fentem and Reece, "A Preliminary Survey of Agriculture."

[72] Mary V. Harris (ed.), "The Autobiography of Benjamin Franklin Harris," *Transactions of the Illinois State Historical Society for the Year 1923*, 89.

[73] *Seventh Census*, 1850, pp. 670-82; Gould, "Wanderings in the West in 1839," *Indiana Magazine of History* (March, 1934), 30:71; Carlson, *The Illinois Military Tract*, 29.

became cheaper when rail contact was made with Wisconsin and Michigan. Wire fencing had a good foothold in Illinois by 1857; and Osage-hedge fencing, which made some headway in Illinois in the 1840's, assumed the proportions of a craze in the fifties.[74]

In the early 1840's, "Frank" Harris of Sangamon County would buy 100 or 200 cattle, feed them during the winter on shock corn purchased for 10 cents a bushel, and by the last of April turn them out on the prairies, where "they had a range of 10,000 acres." An English visitor had to use field glasses to see them. Two men stayed with them during the day; at night they were herded into a stalk field or pound. During the winter Harris bought other cattle in small lots from farmers, to be delivered April 1. In 1853-1854 he bought 950 cattle and 63,000 bushels of corn to add to his own corn for stall feeding.[75]

Between New Berlin and Springfield was prairie, but west of New Berlin were the corn-and-grass farms of James Brown, Jacob and James Strawn, and others. Brown, especially famous as a cattle breeder, had on his 2,250-acre Island Grove farm in Sangamon County in 1855 some 480 acres of corn, 50 cows, 500 steers, and 400 stall-fed beeves. Jacob Strawn, who had come to Morgan County from Ohio about 1830, was doing stall feeding, too; in 1854 he had in corn 2,300 of his 18,000 acres in Morgan and Sangamon counties. A real "cattleman," he was in the saddle day and night. Like Abraham Lincoln, who was his lawyer, Strawn was a freesoil Whig. His ancestry was Pennsylvania Dutch, but his father had been expelled from the Quaker church for shouldering a musket during a Revolutionary War battle. Jacob, too, was a man of action; once when the St. Louis butchers united to fix the price they would pay for his beeves, he broke the conspiracy by sending out men to buy every drove on its way to St. Louis.[76]

[74] Richard Bardolph, "Illinois Agriculture in Transition, 1820-1870," *Journal of Illinois State Historical Society* (September, 1948), 41:254-58.
[75] Harris, "Autobiography," 77.
[76] Harris, 88; Clarence P. McClelland, "Jacob Strawn and John T. Alexander, Central Illinois Stockmen," *Journal of the Illinois State Historical Society* (June, 1941), 34:180-89.

Also in Morgan County was John T. Alexander, who made his first land purchase in 1848, about ten miles east of Jacksonville. At the close of the Civil War he held 7,200 acres in the Sangamon Valley and then added 26,000 acres known as the Sullivant farm or "Broadlands" in Champaign County. He shipped cattle directly from Logansport, Indiana, via rail and water to New York and Boston.[77]

In the midfifties a mania for heavy cattle seized some Illinois cattlemen. Frank Harris in the spring of 1854 shipped on the Illinois Central Railroad 100 cattle averaging 1,965½ pounds gross. His neighbors, Calif and Jacoby, thereupon decided to try to outdo him, and challenged him to a contest. Their 100 head weighed in on the scales next spring at an average of 2,113, Harris' at 2,377. In 1857 Harris produced 100 head averaging 2,786¼ pounds! Twelve of these steers, decorated with tricolored ribbon, were paraded through the principal streets of Chicago, preceded by a marching band and followed by 100 butchers, mounted and uniformed. Meanwhile, however, in 1855 Calif and Jacoby had discovered that the New York butchers took a dim view of such mammoth brutes and refused to pay a worthwhile premium per pound for them.[78]

When, therefore, in the period 1856-1858, cattle prices sagged in the New York market, and Illinoisans found it no longer profitable to send inferior cattle there, they were prepared to send well-finished grass cattle or corn-fed beeves. These they sent by railroad, offering severe competition for the Scioto feeders.

The cattle kingdom of the Ohio Valley, as it existed in the thirty years before 1860, with its complex and changing skein of feeding areas and ranges, had a significance even beyond the utilization of natural resources, the sorting out or classification of the lands, and the specialization of agriculture. Just as the golden touch of the breeder had created a proud tradition in the Ohio Valley, so the ultimately successful grazing, fattening,

[77] McClelland, "Strawn and Alexander," 203; Helen M. Cavanagh, *Funk of Funk's Grove* (Bloomington, Illinois, 1952), 59.
[78] Harris, "Autobiography," 89-92; *New York Tribune,* May 3, 1855.

and marketing of cattle created self-confidence, wealth, and stability.

As in Illinois, feeding advanced into those ranges that had a favorable natural geography, although in east-central Illinois cash grains eventually became more important than feeding. In the ranges on the northern fringe of the Ohio Valley, the Corn Belt was born well before 1860. Much of the Kentucky range, however, has never gone in for feeding; although some corn is grown in the bottoms along the Green River, a Kentuckian trying to describe a skinny woman will still say, "She's as bony as the hips of a Green River cow."[79] The historic feeding regions of the Scioto and the Kentucky Bluegrass, though still proud of their traditions, have since the 1860's become overshadowed in production by the great Corn Belt stretching northwest through Iowa. Today, some of the cattlemen in the Kentucky Bluegrass find it more feasible to buy the corn they use for feeding (it is trucked in from Ohio) than to raise it themselves.

In the Ohio Valley beef empire, corn and cattle were not the only agricultural produce. Eventually, after 1860, the raising of tobacco surpassed beef production in being characteristic of the Ohio Valley as a geographical region; and while corn production has remained important in the valley, the fattening of cattle has become more characteristic of the Corn Belt, particularly the western part. Thus, beef cattle have always been raised in association with—perhaps in rivalry with—other agricultural produce. The proportion of the associations and the outcome of the rivalries were determined by such factors as prices and competition, weather and diseases, soil, labor and technology, the reputation of the local product, the need for capital, the availability of capital, and tradition.

Although some of the pioneers brought cattle with them on flatboats—some 2,500 head accompanied pioneers on flatboats bound for Kentucky during an eleven-month period—it was unusual for anyone, about 1800, to engage in the cattle business

---

[79] Writers' Project of the Works Progress Administration, *Kentucky, A Guide to the Bluegrass State* (New York, 1939), 294.

during his first year breaking the wilderness.[80] Instead, the settler's cattle might consist of an ox for plowing and a milk cow. That first spring the pioneer generally planted one to three acres of corn, with squashes and pumpkins between the rows, and one-half to one acre of potatoes and turnips. If a Yankee, and a patient Yankee, he probably set out a few apple seeds too. The next year wheat, hemp, and flax were added to the crops, and perhaps hogs and a slave (in Kentucky) or hired hand were obtained.[81] Not until the third year, probably, was there sufficient capital to enter the beef-cattle business—and sometimes not until the tenth or twentieth year.[82]

About 1815, the development of the linseed-oil industry gave flax growing a big boost in the Ohio Valley, and Piqua, Ohio, became the largest producer of this commodity in the country.[83] Any possibility, however, that flax growing would rival the beef-cattle industry in the Ohio Valley was removed when the cottonseed oil of the South began to displace linseed oil.

The root crops were, like flax, pioneer crops, the potentialities of which were not limited to the crude clearing in the forest. Turnips and sugar beets had particular significance for the cattle industry, because both could be used as cattle feed, as in Germany. Mindful of the two years of summer drought that had just passed, Henry Clay, Jr., in the fall of 1839 urged his fellow Kentuckians to raise turnips as a supplemental feed for cattle, which, he said, would eat both the tops and the bulbs, a full-grown steer eating two or three bushels of bulbs per day plus straw or hay. A Maryland agricultural paper observed that a single acre of sugar beets should yield at least 1,089 bushels, which, at the rate of a bushel per day to each

[80] Charles Wayland Towne and Edward Norris Wentworth, *Cattle and Men* (Norman, Oklahoma, 1955), 217.
[81] Morrow Scrapbook.
[82] This may be concluded from the fact that the beef-cattle industry in the Miami Valley remained in its infancy for three or four decades after the first of the settlers came in. In most places, of course, the rise of the beef-cattle industry was given impetus by the arrival of what Timothy Flint described as the "second wave" of settlers, men who were not pioneers but men of capital.
[83] W. A. Lloyd, J. I. Falconer, and C. E. Thorne, *Agriculture of Ohio* (Wooster, Ohio, 1918), 54.

steer and cow, would feed seven cattle from December through April.[84]

A Kentucky agricultural paper replied, however, that "in our western country it is cheaper to raise corn to feed and fatten our cattle . . . , than vegetables, and not one tenth the labor and expense." To feed vegetables to cattle, one would have to have barns and stalls. "Six or eight plough boys will make corn enough in the summer, while the stock is on grass, to fatten one hundred beeves, during the winter." It was argued against raising turnips that they "are the most precarious crop of any we attempt to make." Another objection, of course, was that the high water content of roots tends to bloat the cattle with water instead of putting flesh on them.[85] The usefulness of the roots to cattle farmers seems chiefly to have been as a fall crop, planted when it looked as though the corn crop would be short, and fed to the cattle in conjunction with corn.

Although it was pointed out to cattlemen that apples were useful as slop for cattle, the raising of orchards remained unimportant in the Ohio Valley.[86] One day in Illinois, Solon Robinson tells us, a New Englander and a Southerner fell to discussing this neglect of orchard-raising: "New Englander: 'And why don't you set an orchard.' Southerner: 'Well, I reckon may be I will some day—did set out a few trees once, and they grew powerfully, but the cattle soon destroyed 'em.' New Englander: 'And no wonder, for they were set in "the big field," the eternal corn field'."[87]

Dairying might be supposed a natural associate to the beef-cattle industry in the Ohio Valley, especially since the Durham Shorthorn, then the preferred breed in the valley, was a "triple-purpose" breed. Actually, however, dairying attained no more prominence than orcharding. The Southerners tended to be

---

[84] *Franklin Farmer*, December 2, 1837; December 7, 1839.

[85] *Woodford Farmer*, quoted in *Franklin Farmer*, December 16, 1837; *Franklin Farmer*, February 3, 1838; L. H. Bailey, *Cyclopedia of Farm Crops* (New York, 1922), 295.

[86] *Franklin Farmer*, October 21, 1837.

[87] Quoted in Richard Lyle Power, *Planting Corn Belt Culture* (Indianapolis, 1953), 97.

indifferent toward dairying. The production of dairy items required too much skilled labor, it was thought, and hence would not pay. For example, the uniform quality which the export market demanded in cheese seemed difficult to attain, though it was accomplished to some degree after 1841.[88] Many of the cattlemen could not possibly envision themselves tending "milch cows," traditionally the task of the farmwife and children. These men agreed with Nat Sawyer of Madison County, Ohio, who offered for sale in 1836 his dairy herd of 60 cows and three English-stock bulls "because, he finds it difficult to carry... [dairying] on upon a farm with a numerous stock of large cattle."[89] The reputation of Ohio Valley dairy products, moreover, was spotty.

The real competition to corn was being offered by three great crops of the Ohio Valley: hemp, wheat, and tobacco. Sometimes one or the other of these tended to displace corn; other times to be displaced by corn; yet still other times to be combined with cattle farming and even corn farming to mutual advantage.

Hemp affords an example of such combination. During the 1820's hemp partially displaced tobacco in the Kentucky Bluegrass, and grazing expanded. Bluegrass was sown in the woodlands, and rotations of corn, hemp, rye, and clover were begun in the fields. The hemp contributed little to soil exhaustion and helped prevent erosion.[90]

The hemp crop did require cheap labor, however, because the pulling, curing, stripping, stacking, breaking, and rotting took much work. Kentucky had such cheap labor—in fact, eventually a surplus of slaves. Cattle farming could not alone keep this labor force busy, since not all the cattlemen were as fastidious as Scott and Breckenridge about keeping their pas-

[88] Otis K. Rice, "Importations of Cattle into Kentucky, 1785-1860," *Register of the Kentucky Historical Society* (January, 1951), 49:45; Eric E. Lampard, "The Rise of the Dairy Industry in Wisconsin: A Study of Agricultural Change in the Midwest, 1820-1920" (dissertation, University of Wisconsin, 1955), 89, 91-92, 97-99.
[89] *Scioto Gazette*, December 7, 1836.
[90] Rice, "Importations," 40.

tures weeded. Every great hemp county also had large numbers of beef cattle: Bourbon, Fayette, Garrard, Shelby, Woodford, Lincoln, Mason, and Scott. Not every great cattle county, however, had a large production of hemp—Clark and Madison, for example.[91] Obviously, some cattlemen, and not alone abolitionist cattlemen like Madison County's Cassius Clay, shunned a cattle-hemp plantation worked by slaves, even though twenty slaves would ordinarily be enough to do the job. Cattleman Nat Hart of Woodford County explained in 1833: "For several years I turned my attention to the raising of hemp, and succeeded very well in it; but being in the possession of a considerable tract of land well adapted to grazing, and finding that to extend the raising of hemp, so as to make it an object, in my situation, would require an increase of a description of labourers [slaves] that I was unwilling to be taxed with, I declined the culture of it as a leading crop, and turned my attention chiefly to grazing."[92]

As the reputation of Ohio Valley hemp improved, the local farmers began to regard it as a reasonably good cash crop. Nat Hart, apparently regretting his deemphasis of hemp, wrote his daughter in 1841 that "I have no doubt but that it [hemp] is a better business than raising stock at this time."[93]

One Bourbon County farmer proposed in 1842 that his once wealthy county lift itself out of its six-year depression by turning from beef and pork to hemp. Although the South, he charged, had been deliberately resisting (presumably by its low-tariff policy) the rope and bagging industry of Kentucky, he saw hope in the Whig Congress and in the possibility of steam-rotting hemp, which had been successfully done in a neighboring county. He concluded, "The land of Bourbon is emphatically a hemp soil."[94]

Central Kentucky was spared the ravages of a one-crop corn economy. Farmers planted cover crops to protect the exposed

---

[91] *Seventh Census*, 1850, pp. 626-28.
[92] Hopkins, *Hemp Industry*, 3, 19-20, 24, 25-26, 27-28.
[93] Nat Hart to Virginia Shelby, January 4, 1841, Filson Club.
[94] *Western Citizen*, May 6, 1842.

soil in cornfields from winter rains. Livestock returned fertilizer to the earth. Hemp, luxuriant but not voracious like corn, was kind to the soil. Hence that area retained a high degree of fertility long after less protected lands in some other regions had become exhausted.[95]

Wheat, with all its ups and downs in the period 1783-1860, failed to achieve dominance in the Ohio Valley. Wheat straw, a byproduct, constituted a desirable part of the winter feed of cattle, and was regarded as a substitute for grass: "It is maintained, by practical men, that grounds under good tillage will yield as much cattle-food, in roots, straw, etc., as the same grounds would yield in grass, thus leaving the grain as extra profit."[96]

Corn, not wheat, was king. From the first it rated in the West as a "sure" crop; it was unthinkable that any farmer could live without it. By "sure" crop, no one meant that it did not have bad years; the point was that the corn crop, in striking contrast to wheat, never varied more than 25 percent in a single year. The cornfields in early summer were a sight to behold, and the prowess of the corn plant became legendary. A letter from the Muskingum Valley in 1788 reported "that they drove a stake into a cornhill and measured the corn, and that in 24 hours it grew 9½ inches." Besides the usual gourd-corn brought in by the Southerners, men like cattleman Robert Scott tried other varieties too: Manaan, barley-corn, small yellow Canada, Rare Rice.[97] Most of the corn raised in the Ohio Valley was fed to livestock, as we have seen.

Corn was voracious; Buel warned that it was a "gross feeder," leaving little upon the soil for what it took from it. Good farmers recognized the need for crop rotations that would prevent soil exhaustion. Mention has been made of rotations in the Kentucky Bluegrass. A New Englander wrote: "Good husbandry . . . enjoins that culmiferous [small grains, corn, tobacco] and leguminous crops should follow each other in

[95] Hopkins, *Hemp Industry*, 5.
[96] Buel, *The Farmer's Companion*, 157.
[97] Power, *Planting Corn Belt Culture*, 151, 152; Leavitt, "The Livestock Industry," 83; Scott Farm Daybook.

succession, except where grass is made to intervene." However, added Buel, "Indian corn may, under certain contingencies, be made to follow a small grain crop to advantage. . . . Where Indian corn is to succeed small grains, we venture to recommend the clover with the small-grain crop." William Henry Harrison, a stanch rotationist, disagreed that the corn and small-grain rotation would suffice: "Many [farmers] suppose that if they change the crop from corn to small grain . . . that they may then go back again to corn without injuring the land. This is a mistake. . . . A crop of clover succeeding three successive grain crops, and remaining two years, will effectually answer the purpose." Kentuckian Sam Martin went even further, proposing an eight-year rotation plan of two years of hemp, two of corn, one of oats, one of wheat or rye, and two of grass.[98]

Corn was king, but that did not necessarily make the steer prime minister. The corn could be converted into whisky or pork. In the Grand Prairie of Indiana cattle baron Adams Earl owned several distilleries and fed the slop to his livestock.[99] No farmer, so far as I know, became rich by selling corn to distilleries; but this could be said for it—it eliminated the risks involved in feeding the corn to livestock and then marketing the livestock.

The second alternative, the conversion of corn into pork, gained early and widespread acceptance in the Ohio Valley. Hogs were easier to slaughter with assembly-line methods than cattle, and pork packed better than beef. Hogs had big litters and perhaps were easier to take care of than cattle. Most farmers believed that hogs give a bigger return than cattle, arguing that hogs are more efficient feeders than cattle. Hogs were popular, too, for scavenging or cleaning up after cattle; in fact, they would become fat on cattle droppings alone, even where the cattle had not been on corn. The beef-cattle business offered big risks as well as the chance of big profits;

[98] Buel, *The Farmer's Companion*, 156-57, 198; *Franklin Farmer*, November 25, 1837; *Maysville Eagle*, May 14, 1845.
[99] Gates, "Hoosier Cattle Kings," 13.

therefore, most corn farmers who fattened beef cattle hedged by having swine too. While they did so, they still preferred to think of themselves as cattlemen, or cattle feeders, rather than as the much-despised "hawg farmers." Considering that hogs more than held their own in association with beef cattle in the poorer areas of the Ohio Valley, we would have difficulty saying whether the Ohio Valley was more a beef-cattle empire than a hog empire.

Thus, Ohio Valley agriculture was specializing in a corn and livestock economy. Men with capital who might have gone into fruit growing or dairying shunned these activities because of lack of experience in them, because these activities were thought to require too much skilled labor, or because spoilage encumbered both these activities with even greater risks than the beef-cattle business involved. The production of hemp was risky because of the problems of rotting and of foreign competition, and it was left almost exclusively to the slave labor of Kentucky, where hemp gained popularity and helped conserve the soil. Wheat enjoyed two spells of popularity—one before 1819, the other from 1850 to 1857—but otherwise retreated into special areas. Tobacco, a cash crop that required little capital, became characteristic of the poorer areas, though some was produced in the Kentucky Bluegrass and in Montgomery County, Ohio. Beef and pork maintained their hegemony in the Ohio Valley for more than half a century.

Chapter 4

Importations, 1832-1857

THE PERIOD OF THE CATTLE KINGDOM WAS CHARACTERIZED BY importations of Shorthorn stock direct from Britain to the Ohio Valley. This era was divided into two phases (1832-1840, 1850-1857) by the depression of the forties. During the first phase, although almost all the animals being imported to the Ohio Valley were Shorthorns, the merits and demerits of all breeds, including the Shorthorn, were freely discussed with open mind; and furthermore, the mixing of bloodlines and the breeding of partbloods were regarded as sensible, especially in Kentucky. In the late 1830's, however, the controversy about the Sanders importation began, and around 1839 the Longhorn panic struck Kentucky and persisted through the 1840's. When the second phase of the era of importations opened about 1850, some of the breeders insisted that the pedigreed Shorthorns now being imported be segregated from the Sanders stock. A few breeders, especially in Kentucky, still championed the Sanders stock and even refused to give first importance to pedigrees. Partbloods remained in good demand on ordinary farms.

Besides the general reasons operating in the country at large for upbreeding, importations were a logical development in the Ohio Valley in the early 1830's. Prosperous times, and the fact that Ohio as well as Kentucky had now made a deep penetration into the eastern cattle markets where better cattle were needed, made money available to the Ohio Valley cattlemen for expensive importations. The states of Kentucky and Ohio, being situated in what was now the "Old West," where frontier conditions no longer prevailed, enjoyed a substantial agricultural economy. The cattlemen and their families were riding in buggies to the county fairs to see opulent displays of livestock.[1] The Kentucky cattlemen who were buying expensive breeding stock began to build cattle barns to shelter them.

In 1833 John Dunn, one of an energetic family of Scottish immigrants, journeyed from Kentucky to Britain, and while there, he investigated the possibility of buying breeding cattle. He was advised by one correspondent in Scotland to get Ayrshire instead of Holderness, but John decided to buy Shorthorns —six cows and one bull. The animals (or some of them) had pedigrees, but the bulls listed in certain of these pedigrees could not be found in any of the English herd books. The shipment was sent to Kentucky consigned to John's brother, Walter Angus Dunn, whose name the importation bears. John shortly moved to Chillicothe, Ohio, where he became a merchant; Walter already was engaging in enormous land speculations in the west-central Ohio upland and was probably the biggest land speculator in all the Ohio Valley at this time. Breeding cattle from this shipment found their way to both sides of the Ohio River, but later were subjected to attacks similar to those made upon the Seventeens.[2]

Meanwhile, in July of the same year, 1833, Felix Renick drove a herd of Missouri stockers into Chillicothe. It was hot, the drive had been a long one, he was in poor health, and he was glad to be back home in the Scioto Valley. Almost immediately he picked up a rumor—probably at the cattlemen's rendezvous, the Cross Keys Tavern—of the Dunn brothers' activity. Being a man of imagination, Felix proposed that the Ohio cattlemen organize to make an importation of their own, an importation that would dwarf every importation heretofore made in the Ohio Valley; a company should be formed, and he himself would go to Britain next spring. Walter Angus Dunn was his friend and would be able to give him some advice, and Henry Clay, a personal friend of the Renicks, would be able to get doors opened in Britain. By winter the Ohio Com-

[1] William Renick, *Memoirs, Correspondence, and Reminiscences* (Circleville, Ohio, 1880), 31.
[2] Samuel Gutzell to John Dunn, April 20, 1833, in possession of the Ross County Historical Society, Chillicothe, Ohio; Alvin H. Sanders, *Shorthorn Cattle* (Chicago, 1900), 186-87.

pany for Importing English Cattle was organized, with Felix and George Renick and former Governors Allen Trimble and Duncan McArthur playing leading roles; some of the stockholders, however, were not cattlemen but simply investors, such as a Columbus lawyer, Lyne Starling.[3]

Almost immediately a difference of opinion arose as to which breed to import. Henry Clay, who the year before had tried unsuccessfully to "veto Jackson," was now trying to veto any predilection for Shorthorns. He liked Herefords, which he himself had imported in 1816, and now as then this breed came in for animated discussion. That the Herefords were poor milkers was a serious consideration for some Ohio Valley farmers, but not for the big cattlemen. Herefords seemed to be hardier than Shorthorns, but this suggested that perhaps on good pastures and with plenty of corn the steers could not do as well as Shorthorns. Finding opposition to Herefords, Clay tried to persuade the company to purchase Devons. In the end it was decided to let the agents—Felix, his nephew Josiah, and Edwin Harness —make the selections on the basis of what they saw and learned in England.[4]

Equipped with a letter of introduction from Clay,[5] Felix and his companions started on the trip to England, stopping on the way to inspect the Powell herd of Shorthorns near Philadelphia. In England the party visited one breeder after another, including Thomas Bates and Jonas Whitaker, the latter of whom the Americans later selected as their agent. As the group traveled through England, one English observer described it as a "procession stirred up by a few Yankees." Nineteen head of pedigreed cattle—Shorthorns, not Herefords or Devons—were selected.[6] Harness embarked first with eight of the cattle and

[3] Charles Sumner Plumb, "Felix Renick, Pioneer," *Ohio Archaeological and Historical Quarterly* (January, 1924), 33:27; A. Sanders, *Shorthorn Cattle*, 211-12.

[4] Charles T. Leavitt, "Attempts to Improve Cattle Breeds in the United States, 1790-1860," *Agricultural History* (April, 1933), 7:59, 63-64.

[5] Henry Clay to the Honorable Mr. Coke of Norfolk, February 6, 1834, in possession of Joseph V. Vanmeter of Chillicothe, Ohio.

[6] Plumb, "Felix Renick," 28-34; Ohio State Board of Agriculture, *Twelfth Report . . . 1857*, 301-302.

arrived at Chillicothe about July 25, 1834. Felix and Josiah Renick landed at New York on July 20 with the rest.[7]

In the late summer of 1835, a small consignment of seven head was forwarded from England to Ohio. From the consignment the red-roan heifer Harriet had been designated for Felix's son-in-law, James Renick; James' father, Abram Renick of Clark County, Kentucky, later used her (as well as Josephine and Illustrious) in crossing upon his Rose of Sharon family. The heifer calf Josephine (who later dropped Young Phyllis in this country) was consigned to Felix Renick; when in England the year before, he had admired her dam and had been promised the first heifer calf. The bull Skipton Bridge went as a gift from the English breeder Bates to the Episcopalian bishop of Ohio at Kenyon College, in Knox County; and the heifer Hon. Miss Barrington as a gift to the bishop's wife. These two gift animals were later placed in the hands of a Licking County husbandman for breeding.[8]

Meanwhile, the Ohio Company was crossing its imported stock with what improved stock was already in the Scioto Valley. Thus, halfbreeds (which could not have been more than six months old) were exhibited at the third annual fair of the Ross County Agricultural Society in November of 1835.[9]

The final, and largest, importation made by the Ohio Company occurred in 1836—thirty-five head selected by agent Whitaker. Like the 1835 importation, the cattle landed at New York and were sent on to Ohio.[10]

By now the company had such a large stock of English breeding cattle that it was ready for a public auction. On October 29, 1836, cattlemen assembled in the open field lying between Felix Renick's farmhouse and the Columbus-Portsmouth pike. Of the cattle sold, eighteen went to Ross County, eleven to Pickaway, four to Highland, two to Fayette, one each

[7] *Scioto Gazette and Independent Whig* (Chillicothe, Ohio), July 30, 1834.
[8] A. Sanders, *Shorthorn Cattle*, 204-205; Plumb, "Felix Renick," 41.
[9] *Scioto Gazette*, November 4, 1835.
[10] Plumb, "Felix Renick," 41-49. Plumb, a professor of animal husbandry at Ohio State University, had the use of some Felix Renick papers that disappeared after Plumb's death.

to Madison, Pike, and Scioto, and four to Kentucky—an indication of the wide attention attracted by the auction. The cattle sold at high prices ($400 to $2,225), and some of them changed owners immediately, at even higher prices. Joseph Harness bought Young Mary and her heifer calf Pocahontas (sired by Comet Halley); that same season, two Kentuckians, Solomon Van Meter and Isaac Cunningham, purchased a half interest in the two heifers, and after Harness' death they took them to Kentucky.[11] After the October sale the Ohio Company stockholders gave Felix a bull calf, Paragon of the West, as an expression of appreciation for his services.

But now the directors of the company became apprehensive that its publicity and success would give other people the idea of making importations. In view of this probable competition, they decided to liquidate the company. Accordingly, on October 24, 1837, a dispersal sale was held, well attended by spirited bidders. M. L. Sullivant and Company of Franklin County bought the bull Acmon for $2,500; the bull Comet Halley brought an equal price.

The significance of the company's activities may be summed up in four words: precedent, breed, quality, and quantity. The importations were the first made directly to Ohio or any of the territory north of the Ohio River. A trend had already been running toward Shorthorns, but not strong enough that the company—because of the time and size of its importations—could not have turned it in favor of Devons or perhaps Herefords. The quality of the importations was outstanding; although the males included the great bull Comet Halley, perhaps the females proved equally illustrious, giving the Ohio Valley its Young Marys, Young Phyllises and Roses of Sharon. Most of the Shorthorns eventually bred in America were of lines already renowned in Britain before crossing the water; but the Young Marys, one of the greatest families in the United States, were an exception, for they had little or no reputation when

---

[11] Charles S. Plumb, *Little Sketches of Famous Beef Cattle* (Columbus, Ohio, 1904), 51.

the first was taken across the Atlantic.[12] The quality of the Rose of Sharon females is suggested by the fact that when Rose I was eventually butchered, she "killed 71 percent" (contained 71 percent usable beef) in contrast to the usual 61 or 62 percent for prime cattle.[13] Rose of Sharon's daughter Lady of the Lake, purchased by Richard Seymour of Pickaway County, proved to be a great breeder. She never grew into a large cow, but was neat, with a handsome head and prominent eyes. She was sold to George Renick, for whom she dropped five heifers, one, Rose of Sharon II, by Comet Halley himself; all of these heifers left a valuable progeny, some of which were taken by Abram Renick to Kentucky.[14] Finally, much of the significance of the Ohio Company's work lies in the volume of its importations, sixty-one animals in all, and in the fact that because it was a company, the animals became widely distributed. The expectation of the Ohio Company men that competitive companies would be formed proved wrong because a depression intervened, and the example of banding together in a company or syndicate for importing was emulated just once before 1850.

Because of the quality of the cattle and the prosperity of the times, the Ohio Company got high prices at its auctions. The members of the company who were not cattlemen came out best financially. One share of stock bought for $100 paid $300. The cattlemen members bought or kept some of the cattle for themselves, only to find a few years later that they could sell only at ruinous prices because of the depression of the 1840's.[15]

Apparently one or two Ohioans who had purchased cattle from the company in 1836 exaggerated what they had paid for the animals. One Kentucky agricultural paper warned: "An article now [October 21, 1837] going the rounds of the Kentucky press, purporting to render an account of the sale at Chillicothe on the 20th ult., of fifty head of Durham cattle, is

[12] Plumb, *Famous Beef Cattle*, 54.
[13] Ben Douglas Goff, Sr., interview with the author, Winchester, Kentucky, April 11, 1957.
[14] A. Sanders, *Shorthorn Cattle*, 209-10.
[15] Plumb, "Felix Renick," 51-52; A. Sanders, *Shorthorn Cattle*, 211-12.

calculated to deceive the public. It is an account, with some inaccuracies in the prices, of a sale of stock last year. The Ohio Company have advertised a sale of cattle to take place in a few days." A couple of weeks later this paper reported of the Ohio Company's October 24 sale: "the prices far exceed those of their former sale. . . . These purchases are bona fide, and not sham, collusive bargains as some suspicious, incredulous persons declared the first to be."[16]

Relations between most of the Ohio and Kentucky cattlemen, men of known integrity, were good. The Ohio Agricultural Society for its first cattle exhibition appointed two Kentuckians, Cunningham Scott and Abram Renick, to its five-man committee on premiums. There was, moreover, no attempt by the men of the Ohio Company to keep the benefits of their fine breeding cattle from their Kentucky rivals who were willing to pay handsomely; in the late fall of 1837 Duncan McArthur advertised in Kentucky the private sale of the purebred Durham bull Whitaker for $1,200.[17]

A surge of activity in Kentucky started in the final year of the Ohio Company's importations. The bull Otley, which has had few peers in Kentucky, was imported in 1836 from England and sold to Wasson and B. N. Shropshire of Bourbon County for $2,100. Walter Angus Dunn and Samuel Smith joined forces in the same year to import two bulls (one of them being Comet) and four cows. One of the cows proved barren, but to the others the American Adelaide and Mary Ann families trace. In 1836-1837 Henry Clay, Jr., imported eleven head of Shorthorns, apparently unconcerned that the pedigrees were imperfect or missing entirely.[18]

A spirited rivalry arose between the partisans (mostly Ken-

[16] *Franklin Farmer* (Frankfort, Kentucky), October 21, November 4, 1837.
[17] *Franklin Farmer*, November 4, 1837.
[18] Elizabeth Ritter Clotfelter, "The Agricultural History of Bourbon County, Kentucky, prior to 1900" (thesis, University of Kentucky, 1953). 5; A. Sanders, *Shorthorn Cattle*, 187-88, 215. In 1841 Otley sired the cow calf Louan for George Williams. From Louan stemmed many show animals, some of which were sold in Ohio's Miami Valley. William H. Perrin (ed.), *History of Bourbon, Scott, Harrison and Nicholas Counties, Kentucky* (Chicago, 1882), 71.

tuckians) of the Dunn stock and the partisans (mostly Ohioans) of the Ohio Company's stock. The trump card of the Dunnites was the bull Comet; of the company men, the bull Comet Halley. The issue eventually was resolved in the showyard, with Comet Halley besting Comet. Thereupon the company men let the Kentuckians take Comet Halley to the Bluegrass, where Richard Pindell of Lexington had him standing on his farm in 1839 to serve the cows of other Kentucky cattlemen.[19]

The stock of three important breeders, John H. Powell of Pennsylvania and David Sutton and James Garrard of Kentucky, was more or less dispersed in 1837-1838, to the benefit of Kentucky cattlemen. Cunningham and Warwick purchased cattle at the Powell sale in Philadelphia on September 12, 1837, and brought these imported cattle to Kentucky. Two weeks after the Powell sale, David Sutton of Fayette County auctioned off more than fifty pedigreed Durhams, and a Kentucky farm paper crowed that "this stock commanded better prices than did the imported cattle lately sold at Powellton, near Philadelphia." After the death in 1838 of James Garrard II, who lived at Mount Lebanon, three miles north of Paris, much of his stock was dispersed by public sale—an event which one authority calls the beginning of purebred-Shorthorn breeding in Bourbon County. John Pratt returned home to Great Crossings, Scott County, from the Garrard sale and told his neighbors about some other cattle, belonging to Brutus Clay, which he had seen in Bourbon County, and the neighbors were interested in buying.[20] Brutus had purchased Matilda at the Garrard sale, and next year (1839) she dropped a heifer calf by Otley; this calf, hence, represented a mixing of the Seventeen strain with the newer lines. The distribution of full-blooded or imported breeding stock through Kentucky continued. Walter Dunn, Charles S. Clarkson of Cincinnati, and L. H. Shirley of Louis-

[19] A. Sanders, *Shorthorn Cattle*, 193, 205; *Franklin Farmer*, November 23, 1839.
[20] *Franklin Farmer*, October 7, 28, 1837; Clotfelter, "The Agricultural History of Bourbon County," 5; John Pratt to Brutus J. Clay, October 31, 1838, in possession of Cassius M. Clay, Paris, Kentucky.

ville were selling full-blooded Durhams in 1838 and 1839.[21]

But the year 1839 proved remarkable in Kentucky breeding circles chiefly for the unusual number of importations made directly from abroad to the Inner Bluegrass. Reuben Hutchcraft imported seven pedigreed Shorthorns to Bourbon County, which had held its first county fair three years before. Dr. Samuel Martin, Hubbard Taylor, and J. P. Taylor, all of Clark County, made an importation that arrived in Clark the last week in October; some of these cattle were exhibited the next year in the second Clark County fair.[22] The Fayette County Importing Company, the first importing syndicate in Kentucky and the second in the Ohio Valley, was organized in 1839. R. T. Dillard and Nelson Dudley went to England as agents and bought twenty-one head of cows and heifers and seven bulls. Placed on Sutton's farm near Lexington the animals were sold at auction in July of 1840. By this time the agricultural depression had set in, and so the prices, though good for the times, were not as high as at the Ohio Company's sales. All the buyers were from the Inner Bluegrass; eight head went to Fayette County; five each to Bourbon, Scott, and Mercer; and four to Jessamine. None went to Clark, which presumably was being satisfied by the Martin-Taylor importation.[23]

The firm of Shirley and Birch of Louisville in 1839 bought three cows and four bulls from Samuel Waite, who had imported them through New Orleans. The next year Waite made another importation of eight cows and two bulls. Apparently most of the cows in the latter purchase went to Tennessee. As William Warfield of Kentucky later warned of "errors and blunders" in the pedigrees of the 1840 importation, these cattle were possibly the source of some of the deceptions of which the Tennesseans were complaining.[24]

[21] Brutus J. Clay Stock Book, 1830, Cassius M. Clay collection; *Franklin Farmer*, November 23, 1839. Matilda was also served by Comet, whom the Dunns still had in Kentucky; but the calf died.

[22] Lucien Beckner, "Kentucky's Glamorous Shorthorn Age," *Filson Club History Quarterly* (January, 1952), 26:41-42; A. Sanders, *Shorthorn Cattle*, 216-17; *Franklin Farmer*, November 2, 1839.

[23] A. Sanders, *Shorthorn Cattle*, 217-19.   [24] Sanders, 221-22.

By the early 1840's Brutus Clay had the foundation stock for the great breeding herd whose blood was to become so widely disseminated in the Middle West. In 1837 he had acquired Primrose I, the cow that Cassius had bought from James Garrard four years before. She immediately calved Marygold by Oliver, a bull who had Powell blood. Marygold and Matilda, purchased at the Garrard sales in 1838, now became Brutus' foundation cows, Exception his chief bull. Since the Garrard stock that Brutus owned traced back to the Sanders and Patton stock, his defense of the Seventeens must have occasioned no surprise.[25]

In the discussions of the various cattle breeds throughout the Ohio Valley, the Shorthorn emerged as far and away the most popular. Although opponents complained that the Shorthorns were too large, small-boned, and inactive, proponents countered by pointing out their fatting ability and their docility. The most prominent type of Shorthorns was the Durham —heavy, squarish animals originating in the county of Durham, England. A spirited discussion arose over whether Shorthorns or Devons walked the faster—an important question in those days of droving. Although the Devons appear to have had the edge here, "no animal but the horse out travelling them," the proponents of Shorthorns argued that they were no mean travelers either. The growing popularity of the Shorthorn breed in Kentucky was not arrested even by the charge that "in the South"—an important market for Kentucky breeding cattle—"most [Durhams] . . . die of disease before they become acclimated." By 1837, some two years before the Longhorn panic struck Kentucky, some Kentucky breeders were fearing that the popularity of the Shorthorn breed was causing it to be so inbred that it would fail to pass on all of its desirable traits to later bovine generations; the proper course, they thought (not foreseeing the Longhorn panic), was to cross the Shorthorns with other breeds. In 1849, after the Longhorns had fallen into discredit, Lewis Sanders urged that a half-dozen

[25] Brutus J. Clay Stock Book, Cassius M. Clay collection.

young bulls be imported from Holland or northern Germany "at once" to cross with the Kentucky Shorthorns, but nobody would make such an importation.[26]

Indeed, part-blooded Shorthorns enjoyed great popularity in Kentucky. Grades sired by a Shorthorn bull were worth at least $10.00 more per head than scrub calves.[27] But which breed should be crossed with the Shorthorn? One breeder reasoned: "A cross with the Devon or Hereford would sacrifice milking properties: the dwarfish and ill-shaped Alderney, is not to be thought of: the Ayrshire or the Long Horn, possess no excellencies that the Durham does not possess in a greater degree: and the Holderness, and the Teeswater Short Horns, compared with the improved race, are as the crude ore to the manufactured and polished metal. . . . Experience has already proved, that the cross between the Durham and our common race, is a most judicious one." He advised the grading system, with constant resort to a pureblood Shorthorn bull.[28]

By 1839, however, the old Kentucky practice of breeding pureblood bulls to scrub or grade cows was becoming coordinate with a new Kentucky practice of breeding "best" (which did not necessarily mean fullblood) bulls on "thoroughbred" cows. Wrote Dr. Martin: "The breed will extend rapidly by means of part blooded bulls. Three quarter blooded Durham bulls can be bought now for less money than they will bring as beef when cut and fattened." He put little faith in appearance. His "best" bulls he defined as the best breeders; and the best breeders were determined not by appearance but primarily by trial and secondarily by pedigree: "Some animals that are very defective in appearance, if they have fine pedigrees are extraordinary breeders, and I had rather put my fine cows to such, than to the finest animals that are deficient in

---

[26] Leavitt, "Cattle Breeds," 59; *Franklin Farmer*, October 19, November 4, 9, 1839; *Western Citizen* (Paris, Kentucky), September 28, 1849.

[27] Otis K. Rice, "Importations of Cattle into Kentucky, 1785-1860," *Register of the Kentucky Historical Society* (January, 1951), 49:40.

[28] Henry S. Randall to the *Albany Cultivator*, quoted in *Western Farmer* (Cincinnati), March, 1840, pp. 216-18.

pedigree. The next to trial, in determining the breeding value is pedigree."[29]

Here, ironically, was an early origin of the pedigree craze, despite the fact that Martin was advocating partbloods. Obviously, with Kentuckians disagreeing among themselves as to the relative importance of performance, pedigree, and appearance, but with most Kentuckians agreeing on the desirability of mixing bloodlines, the ground was ready for seeds of suspicion to be sown about the purity of Kentucky stock and even the salability of Kentucky beef. The mixing of bloodlines involved not only the native or scrub cattle but also all the importations; thus a Kentucky cattleman described one of his cattle in 1841 as "a yearling male descended from a Cunningham cow . . . a Beautiful roan combining Sutton's stock of the Ohio Comp'y's importation also Powells' stock as well as the importation of '17." In view of this mixture it is astonishing to find him telling his sister that this animal was "in no way related to any of your cattle"; apparently her cattle were uncrossed "Pattons." With bloodlines becoming so mixed, some fine cattle defied classification; the Jessamine County Agricultural Society did not indicate breed or origin in awarding premiums at its 1839 fair.[30]

This long, liberal tradition among Kentucky breeders did not give way entirely when the Seventeens controversy and the Longhorn panic gave rise to the pedigree craze. Some cattlemen were still more convinced by appearance than by pedigree, and pedigrees did not insure high prices for inferior animals.[31]

A brisk trade existed in unpedigreed partblood Shorthorn breeding stock—the type of stock that was within the reach of farmers of modest means. J. Stoddard Johnston speculated on such cattle, both buying and selling within the Bluegrass itself. Occasionally, too, he sold cattle in Indiana. On August 18,

[29] *Franklin Farmer*, November 9, 1839.
[30] William P. Hart to Virginia Shelby, May 5, 1841, in possession of the Filson Club, Louisville, Kentucky; *Franklin Farmer*, November 16, 1839.
[31] J. Stoddard Johnston, "Farm Book and Journal of events more especially bearing on Agriculture" (1855-1865), Filson Club.

1855, he bought from Richard West two heifers with calf, "of fine blood though he had no pedigrees," for $120 each; on the same day from Evan Stevenson, "two heifers not of as fine quality, at $75"; and on August 20 from Lewis Offut, two "very superior, neat heifers" at $100. To these six heifers were added on September 1 thirteen head of two-year-old cattle, twelve steers and one heifer, for which he paid $327; and on September 5, twelve more for $290, "while in addition we bought 9 cows and two old oxen." On September 8 he wrote in his journal: "Smith came over and we sold him the six heifers specified above with the others, in all 8 at $450—being a little over $56 a piece— This is a satisfactory sale as it gives us a gain of $97 in the transaction—"[32]

Thus, while Shorthorns and partblood Shorthorns became the favorite breeding stock, the discussion of the other breeds continued, and some of them actually had a foothold in the Ohio Valley. The Hereford breed seems, however, to have been left behind in the growth of herds in this region. By 1839, some six years after Henry Clay had recommended Herefords to Felix Renick's consideration, Dr. Martin wrote "no breeder in Kentucky with whom I am acquainted would prefer the Herefords to the Short Horns." One authority has estimated that not two hundred Herefords had come into the entire country by 1880. Two English farmers near Elyria, in the Western Reserve of Ohio, imported some Herefords directly from England in 1852-1853, but neither these animals nor their progeny penetrated southward in any number.[33]

Devons made good oxen, did well on poor land, would yield milk with high butter-fat content, and produced good beef. The reputation of Devons for doing well on poorer soils arises from the fact that they are a light breed, but it was thought that this same smallness prevented them from taking fullest advantage of rich feed. Too, their milk yield, though rich, was

[32] Johnston Farm Book.
[33] *Franklin Farmer*, November 9, 1839; Henry W. Vaughan, *Breeds of Livestock in America* (Columbus, Ohio, 1931), 62; Ohio Board, *Report* . . . *1857*, p. 118.

small.[34] The proponents of Devons persistently argued that they walked faster and lost less flesh on the drive than did Shorthorns; one drover in 1841 recommended the crossing of Devons and Shorthorns.[35] The notorious John (Ossawatomie) Brown imported some Devons from England to the Western Reserve of Ohio in 1842, and in the next few years importations were made to Elyria, Pittsfield, and Oberlin; but few if any of these animals or their progeny went to southern Ohio. Some were seen in the Sangamon Valley of Illinois in 1843. Although Devons were introduced from New York to the Miami Valley around 1849, most of the cattlemen of the nearby Scioto Valley had no preference for them. After all was said and done, Devons remained rare in the Ohio Valley, even in poor-soil areas, and by 1860 even the oxen of this breed were losing place, victims of the horse-powered mechanization of agriculture.[36] It should be remembered, however, that Yankees had brought common Devons, the traditional breed of New England, to Marietta before 1790, and that the native or scrub cattle of the Ohio Valley originated at least in part from Devons.

Ayrshire cattle came into Ohio about 1848 and soon gained a substantial foothold because of their milking properties, but the general disinterest which the farmers of the Ohio Valley had for dairying tended to keep the breed from spreading much south of Licking County.[37]

While Kentuckians were importing breeding cattle directly from abroad during the late 1830's, cattlemen of the newer

[34] Leavitt, "Cattle Breeds," 63. One agricultural editor thought that Devons might be advantageous in the poor-soil areas of Kentucky. *Franklin Farmer*, October 19, 1839.

[35] *Franklin Farmer*, October 19, 1839; January 11, 1840; *New Genesee Farmer* (Rochester, New York), April, 1841.

[36] Burkett, "Ohio Agriculture," 112-13; W. C. Flagg, *Agriculture of Illinois (1682-1876)*, Illinois Department of Agriculture, *Transactions* (1875), 13:322; Josiah Morrow Scrapbook, in possession of the Philosophical and Historical Society of Ohio, Cincinnati; Robert Leslie Jones, "The Beef Cattle Industry in Ohio prior to the Civil War," *Ohio Historical Quarterly* (April, July, 1955), 64:317.

[37] Burkett, "Ohio Agriculture," 114-15; Jones, "The Beef Cattle Industry," 311.

states of Indiana and Illinois were trying to accelerate the up-breeding of their stock. On the open range of the Illinois prairie, bulls ran at large. This situation pleased small farmers who could not afford good bulls and therefore were happy to have roaming bulls serve their cows free; but the big cattlemen, arguing that such a situation prevented the efficient control of breeding, secured a law requiring that bulls be kept off the range. Cows, however, were still legally allowed to run at large, and this practice gave rise to the expedient of keeping all calves, whether good or bad, as bait to bring the cows home every night for milking.[38]

Captain James N. Brown, who was to become famous for his Island Grove herd, moved from Kentucky to Illinois in 1833-1834, bringing with him some Shorthorns that may have been the first highly bred animals of that breed to reach the Sangamon or Illinois River valleys; he settled in Sangamon County, just over the line from Morgan. In 1837 one Bradshaw of Edgar County, Illinois, bought three breeding cattle from George N. Sanders of Grass Hills, Bourbon County, Kentucky. Next year, Chris Whitehead of Franklin County, Indiana—hill country early settled by Southerners—made the first importation of Shorthorns directly from England to Indiana. Illinois and Indiana got their breeding stock in these early years more from Kentucky than from Ohio; breeding stock seemed to follow the migration of Kentuckians into Illinois and Indiana, and these removed Kentuckians were at first some of the leaders of upbreeding in the two newer states. Indeed, there was actually some flow of breeding stock from Indiana to Ohio; most of the descendants of the Whitehead importation were bought up by the nearby Shakers of Ohio's Miami Valley. In 1839 James D. Stevenson of Greencastle, Indiana—on the drainage frontier—purchased Colonel, a young purebred bull, from A. D. Offut (of the partnership Marshall, Washington, and Offut) of

[38] Clarence P. McClelland, "Jacob Strawn and John T. Alexander, Central Illinois Stockmen," *Journal of the Illinois State Historical Society* (June, 1941), 34:196; Leavitt, "Cattle Breeds," 57.

Fayette County, Kentucky. General Sol Meredith of Cambridge City, Indiana, which was on the National Road and also on the drainage frontier, made his first purchase of Shorthorns in 1836, and in 1851 he founded his Oakland Farm herd of Shorthorns.[39]

Scattered evidence exists of the presence of highly bred cattle in Illinois during the early 1840's, by which time most importations from abroad to the Ohio Valley had ceased because of the depression. G. W. Flagg of Perry County advertised a Shorthorn bull in 1841. The same year, former Governor Levi Lincoln of Massachusetts sent some crossings of Ayrshire and Shorthorn to his son in Alton; in a few years other Yankees like Edward Sumner and John T. Alexander would take over a large share of the leadership of upbreeding in the Indiana and Illinois portion of the Ohio Valley. James McConnell had Devons near Springfield in 1843. The next year an imported Shorthorn passed through Chicago on its way to Bronson Murray of Lasalle County, and some of its progeny may have been taken southward.[40]

Although the depression starting in 1840 left most Ohio Valley cattlemen with neither the funds nor the optimism necessary for importations, there were exceptions: Robert Aitcheson Alexander and the Dunn family. These Scots had plenty of money. Alexander, of the Kentucky Bluegrass county of Woodford, possessed an inherited fortune; the Dunns had grown wealthy on land speculations carried on in the west-central Ohio upland. Alexander actually made some importations from England during the decade. Walter Angus Dunn, of Lexington, Kentucky, had sent the fullblood but unpedigreed bull Accomodation to Ross County, Ohio, by 1839. When he died, his sons settled upon their inherited lands in Madison County, Ohio. They had rented and traded stock back and

[39] A. Sanders, *Shorthorn Cattle*, 277; McClelland, "Strawn and Alexander," 196; *Franklin Farmer*, October 28, 1837; November 9, 1839; W. C. Latta, *Outline History of Indiana Agriculture* (Lafayette, Indiana, 1938), 190-92.

[40] Flagg, *Agriculture of Illinois*, 322; McClelland, "Strawn and Alexander," 196.

forth with Brutus Clay, and they brought with them a lot of good Shorthorns, including the pride of the Dunnites, the bull Comet, who, despite his defeat by Comet Halley, was used in Ohio until 1845 and proved himself an excellent getter. The Dunns now traded cattle back and forth with their neighbor Ohioan, David Selsor, whose "Christmas beef" became famous in the East.[41]

In the early 1850's others resumed their upbreeding activity. James S. Matson, a Bourbon County cattleman, advertised for sale in 1850 about twenty-five Durham cows and heifers, some of which had young calves, two imported cows, and three bulls four or five months old. The animals, except the two imported cows, were descendants of the importations of the late 1830's. About 1852 Matson imported two famous bulls, John O'Gaunt and Javelin.[42]

A group of Inner Bluegrass cattlemen organized in 1853 the Northern Kentucky Importing Company, in which the cattlemen held stock. The company received wide notice; a Covington man wrote Brutus Clay urging importation of the Westfriesian breed from Holland. Charles T. Garrard of Bourbon County, with Solomon Van Meter and Nelson Dudley of Clark County, went to England as agents. Having landed at Liverpool on March 31, they inspected a herd near there but were not much impressed. Then they went on to Yorkshire, setting up headquarters at Otley. Ten bulls and fifteen heifers were selected. There was a great deal of difficulty in securing ship transport for the animals at a reasonable rate, because British law would not allow livestock to be carried on vessels "carrying passengers." When the animals were disembarked at Philadelphia, "multitudes" crowded the stables to see them. Between Philadelphia and Pittsburgh, Garrard and his companions spent two nights sleeping in a cattle car on a pile of

[41] John Ashton, *History of Shorthorns in Missouri prior to the Civil War*, Missouri State Board of Agriculture, *Monthly Bulletin*, 21:21; John G. Dun to Brutus J. Clay, July 24, 1839, Cassius M. Clay collection; A. Sanders, *Shorthorn Cattle*, 249; David Selsor Graham, interview with the author, Midway, Ohio, September 9, 1955.

[42] Clotfelter, "The Agricultural History of Bourbon County," 6-7.

hay with four of the cattle.⁴³ It was worth it, though; for the herd, which cost $11,780 in England, sold for $47,860 in Kentucky. The company at its auction of August 18, 1853, on Brutus Clay's farm accepted bids from Kentuckians only and required (at Brutus Clay's instigation and over Garrard's protests) bond for double the purchase price of the stock that it would not be taken out of Kentucky for at least a year. The cattle were sold to breeders from Bourbon, Clark, Fayette, Scott, Woodford, Franklin, and Jefferson counties. Brutus Clay took the bull Diamond, who proved impotent.⁴⁴

Matson's activities and the purchases by Bourbon County breeders at the Northern Kentucky Company's sale were not the only evidences of the advanced state of breeding in Bourbon County. The imported bull of the noted Princess family, Lord Vane Tempest, owned by J. M. Sherwood of Auburn, New York, stood at Ben Coleman Bedford's farm. Bourbon County was the first county in Kentucky to have a bull from the famous Bates herd in England. This was Locomotive, purchased in 1854 for $1,800, pedigree guaranteed. George Bedford secured this guarantee because he was worried that the pedigree on the dam's side was not as full as it might be. These Bates Shorthorns differed from the Booth Shorthorns in that they would fatten before four years of age and they had short legs. Locomotive was owned successively by Bourbon Countians James Letton and Brutus Clay. The latter had him standing for heifers in July, 1854. The next year, while visiting his ancestral seat in Scotland, Robert Aitcheson Alexander shipped the bull Sebastopol to Bourbon County, where he was to be owned jointly by Clay, Bedford, Duncan, and Garrard.⁴⁵ Brutus

---

⁴³ O. G. Bargen to Brutus J. Clay, May 15, 1853, Charles T. Garrard to Brutus T. Clay, April 1, July 24, 1853, Cassius M. Clay collection.

⁴⁴ Clotfelter, "The Agricultural History of Bourbon County," 8; Rice, "Importations," 44; Charles T. Garrard to Brutus J. Clay, August 2, 1853, John B. Payne to Brutus J. Clay, January 30, 1854, Cassius M. Clay collection.

⁴⁵ *Western Citizen* (Paris, Kentucky), February 10, 1854; Clotfelter, "The Agricultural History of Bourbon County," 5; George M. Bedford to Brutus J. Clay, February 2, 1854, Brutus J. Clay, Register of Names and Memorandum Book, 1854-1872, R. A. Alexander to Brutus J. Clay, July 20, 1855, Cassius M. Clay collection.

Clay sold breeding stock from his Bourbon County farm to cattlemen in Illinois, the northern Missouri prairie, and Iowa; one man in Van Buren County, Iowa, wrote Clay in 1855, "Some of our farmers are in favor of purchasing in the Scioto Valley, Ohio, but my preferences are toward Old Kentuck." Clay also sent breeding cattle in the 1850's to Virginia's Big Levels, ancestral home of the Kentucky Renicks.[46]

In 1853, the same year the Northern Kentucky Company was selling breeding cattle to Kentuckians, an importing company was organized in Scott County, also in the Bluegrass. Its agents, W. Crockett and James Bagg, bought seven females and five bulls, which were imported to Kentucky in 1854. The men of the Scott County Company now changed their name to the Kentucky Importing Company and made another importation—fifteen cows and six bulls. At the auction Robert Alexander paid $3,500 for the roan two-year-old bull Sirius.[47]

The figures of two men—Robert Aitcheson Alexander (whom we have mentioned in connection with the 1840's) and Abram Renick, both of Kentucky—loom especially large in Ohio Valley breeding during the 1850's and for some years thereafter. Alexander had left his native Scotland because, says a Kentucky story, he was having "woman trouble." He bought some excellent land in Woodford County, and there, instead of building a Georgian or Greek Revival house as the Kentuckians did, he erected a Scottish-style stone house, and stone barns too. A fancy farmer, he had racehorses as well as cattle.[48]

Besides his own importations of 1852 and 1853 and subsequent smaller purchases from abroad, Alexander also acquired some imported cattle from the Northern Kentucky and Scott County companies. But he did not insist on imported cattle; rather, he had an eye for beauty of form—in racers, trotters,

[46] Brutus J. Clay Register, Timothy Day to Brutus J. Clay, June 25, 1855, A. Rodgers to Brutus J. Clay, September 7, 1855, Samuel Lexington to Brutus J. Clay, November 23, 1858, Cassius M. Clay collection.
[47] A. Sanders, *Shorthorn Cattle*, 257-58.
[48] Ben Douglas Goff, Sr., interview with the author, Winchester, Kentucky, November 8, 1955.

cattle, sheep, and women. His cattle herd is said to have been founded on a varied basis, including "Pure Booth, pure Bates, Knightley, Mason, Wiley, Whitaker, 'Seventeen,' every strain nearly that has ever been known on the continent." By 1860 this herd numbered about two hundred head. Alvin Sanders calls it "beyond all question the best collection of Shorthorns then in North America." "Indeed," he adds, "it is doubtful if its superior, size considered, existed at that time in either England or the United States."[49]

Alexander's character, however, was not universally admired by his contemporaries: "A good deal of dissatisfaction was expressed [at the Lexington Fair of 1855] at Alexander's taking so many premiums, though few, I judge, could dissent from the decisions. His bearing towards gentlemen desirous of inspecting his stock at his stalls on the ground was the subject of considerable remark and censure. He does not understand the Kentuckians, I should judge, though he is a Kentuckian now, yet raised in England and holding English citizenship. His wealth is great amounting to an income of $100,000 per annum."[50]

Though successful, Alexander was quite capable of making mistakes. Although he was in Scotland himself in the summer of 1855, Clark County tradition says he hired Benjamin Groom, who lived in a Gothic-towered mansion east of Winchester, to select pedigreed cattle in Scotland for him.[51] Groom came back with what the legend says was a sorry-looking bull, though known by the impressive name, Duke of Airdrie. Alexander, so the story goes, figured he was stuck and that there was nothing he could do about it. Old Abram Renick, however, hearing of this and stirred by curiosity, got on his horse and rode over to see the bull and talk to the Scot. Alexander said he would rent the bull for any price, just to get it

[49] A. Sanders, *Shorthorn Cattle*, 263-70.
[50] Johnston Farm Book.
[51] Groom is remembered by some as a crook who after the Civil War sent cattle with false pedigrees to California. Later he moved to Texas, where he engaged in land speculation, founding the town of Groom. Goff interview.

out of sight. Renick took him up on it and bred the Duke to the Rose of Sharon cow Duchess. The produce was Duke of Airdrie 2478, the bull that was to mean so much to Renick's herd. As it turned out, Alexander's mistake was not in buying the Duke, but in renting him.[52]

The Scot, being very particular about form, had perhaps been unreasonably disappointed in the appearance of Duke of Airdrie; for though his hips were too prominent, the animal had, according to Alvin Sanders, "no serious faults." Yet the bull proved more valuable in performance than in looks. Abram bred him to the fullest extent possible on the female hierarchy that had stemmed from Rose of Sharon of the Ohio Company importation. From him came a line of four more Dukes, of whom Airdrie 2478, when bred on the Rose of Sharon females, became almost the making of Renick's herd.[53]

The assumption has been that "Old Abe" knew animals better than most other breeders did, and that the slightest symptom of any sort about an animal was immediately evident to him. The Bible expresses it, "The eye of the master fatteneth the stock." Abe appears in a different light, however, in William Warfield's report to Brutus Clay: "I much fear that Renick has injured your white heifer [by too much breeding].... She has been straining and carries her tail up."[54]

Keys to Abram Renick's success after 1855 were his rule against selling heifers and his practice of inbreeding. Because Abe would never sell any of his heifers, the notion spread through Kentucky that there was something very special about them. He bred Duke of Airdrie and his progeny on the Rose of Sharon females year after year. Today this inbreeding would not be considered good practice, but at that time it attracted favorable attention. The popularity of Duke of Airdrie in Kentucky aroused an unfounded prejudice, there and elsewhere, in favor of Bates Shorthorns and against Booth Shorthorns—a

[52] Goff interview.
[53] A. Sanders, *Shorthorn Cattle*, 300-301; Plumb, *Little Sketches*, 64.
[54] William Warfield to Brutus J. Clay, August 2, 1855, Cassius M. Clay collection.

prejudice deplored by Alexander, who had financed the importation of the Duke.[55] It is ironical that the Duke, who at first was despised as an ugly duckling, became the most prized sire in Kentucky and revolutionized Shorthorn breeding; while Captain Brown's Rachel, for whom her owner had anticipations to the tune of $3,000, proved a fizzle.

Abe made a deal to purchase from Ben Van Meter the second house that had been built by Matthew Patton. As payment, Van Meter was allowed to pick four cows from Renick's fabulous herd. Van Meter, who, besides suggesting that the Eskimos be captured and enslaved, was fond of drinking bourbon from a jug while watching his horses race on his private track, may have known horses better than cattle; he made his choice of four cows, but presently complained that one was barren. So he decided to prosecute, but had he known his central Kentucky neighbors better, he would not have made the mistake of hiring a fancy-dan lawyer from Louisville. Renick, canny in the ways of central Kentucky, hired a squirrel-headed lawyer from Cynthiana, and the battle was on. The case was argued for three days in the Clark County courthouse, which was jammed to overflowing. After the Louisville lawyer delivered his brilliant summation, the defense lawyer arose and observed: "That was a fine speech. The gentleman from Louisville is a healthy man of 35, married to a fine woman, but how many children does he have? None!" This convinced the jury that form had nothing to do with the matter, and the verdict was awarded to Renick. Van Meter, however, sold the three fertile cows ten years later for far more than the Patton house was worth.[56]

Van Meter could take consolation, too, in the fact that whenever a cow actually proved barren or a bull impotent, men at first were inclined to deny the inadequacy in the animal and to

[55] Goff interview; A. Sanders, *Shorthorn Cattle*, 287-96, 302. Actually, this rule was not a new idea. The farmers' maxim, "Keep the she-stuff," dates back to Biblical lore.

[56] Beckner, notes prepared for the author and copied at the Filson Club, Louisville, Kentucky, October 20, 1955; Goff interview.

blame the keeper instead. Thus, one cattleman blamed "the nigger who attends to 'Diamond'" for that bull's failure with the cows.[57]

Most contemporaries of bewhiskered, gaunt-faced "Old Abe" regarded him with great respect as one of the wisest and most successful cattle breeders in America. In spite of his success, he lived modestly, with a plebeian fondness for plug chewing tobacco. Unlike most Kentuckians, he was well enough off that he could, and did, sit on his front porch, feet propped up, chewing tobacco and watching his Rose of Sharon herd graze. One day, so the story goes, a carriage drove up and a lackey came to Abe's door, announcing that his master, a titled Englishman, wished to buy some of Abram's cattle; replied Abe, with a spit of tobacco juice, "Tell your master that he doesn't have enough money to buy any of my cattle."[58]

The first male offspring of Abram Renick's Duke of Airdrie became the property of George M. Bedford of Bourbon County, who apparently could afford it. This was Bell Duke of Airdrie, who was later to win many prizes, including the $1,000 sweepstakes at St. Louis in 1858 and the $600 championship there in 1860.[59] Bedford became a strong proponent of the short-legged Bates cattle, which swept to popularity with the end of long-distance droving.

Cattlemen north of the Ohio River were active, too, in upbreeding and importing during the years 1851-1854. Thomas Wilhoit of Henry County, Indiana—on the drainage frontier—began breeding Shorthorns, at first unrecorded, in 1851. His neighbors, who had no idea of the prices improved breeding stock might bring, laughed at him for his extravagence in paying $35.00 per head for these cattle, but Wilhoit persisted in attempting upbreeding and eventually won a reputation. In 1853 Dr. A. C. Stevenson of Putnam County—on the southern margin of the Wabash Valley belt of corn farms—made an

[57] John B. Payne to Brutus J. Clay, January 30, 1854, Cassius M. Clay collection.
[58] Goff interview.
[59] Clotfelter, "The Agricultural History of Bourbon County," 10.

importation of Shorthorns directly from England to Indiana, the first since Whitehead's fifteen years before.[60]

Ohioans, meanwhile, tried to revive the leadership once provided by the now defunct Ohio Company for Importing English Cattle. The Scioto Valley Importing Company was formed in 1852, and the veteran Dr. Watts and George W. Renick (son of the deceased Felix Renick) were appointed to go to Britain. There they purchased ten bulls and seven females that were sold at auction at Watts' farm near Chillicothe. Two bulls brought substantially over $2,000 each. All of the cattle went to Ohioans, about half of whom lived not in the Scioto Valley but in the west-central Ohio upland. The Madison County (Ohio) Company, organized a year after the Scioto Company, imported fifteen bulls and nine cows. Only one of the bulls and none of the cows went outside of the upland, which obviously was coming into its own as a breeding center. Although Alexander Waddle of South Charleston had already denounced Kentucky stock of Sanders' origin, some of the upland cattlemen were bringing in this stock.[61]

Robert Corwin, brother of the Whig politician Thomas Corwin, made an importation of Scottish Shorthorn breeding stock to the Miami Valley in 1854.[62] Many of these animals went to the Shaker community of Union Village. Corwin and Alexander were the first to bring Bates cattle from Scotland, the breed itself actually having originated in Northumberland.

A friend of Brutus Clay bought from E. A. Robinson some cattle which were represented as being of the Corwin importation. But they were not. Corwin wrote Clay that he had taken three of the cattle he had imported and the Union Village Shakers the rest, and that neither had sold any. "I should be very sorry if any Kentucky breeder has been imposed upon in Ohio, for we have been denouncing some of your people for

[60] Latta, *Outline History*, 190, 192.
[61] A. Sanders, *Shorthorn Cattle*, 249-54; *Cincinnati Gazette*, quoted in *Ohio Cultivator*, August 15, 1852; James Fullington to Brutus J. Clay, n.d., Cassius M. Clay collection.
[62] John C. Hover and others (eds.), *Memoirs of the Miami Valley* (Chicago, 1919), 2:352.

imposing on us," concluded Corwin, perhaps recalling Governor Trimble's complaints of 1838. Finally, in 1859, Corwin offered his entire herd for sale, "though just now it is not in high condition."[63]

What the Corwin importation did for the Union Village Shakers caused eyes to pop here and there in Warren County, where a few bigots had periodically proposed to drive the Shaker community out of the county. Farmers who had scoffed at Shorthorns now observed their Shaker neighbors selling yearling steers with evidence of Shorthorn blood for up to twice as much as their own scrubs would command. The first year (1855) after the Corwin importation, the Union Village Shakers sold $8,420 worth of blooded Durhams. Next year, 1856, their Cap't. Balco was reported valued at $10,000 and was standing for $100 a cow. The price of the Shakers' purebred calves ranged from $100 to $1,000, according to quality and pedigree.[64]

The Clark County (Ohio) Company was formed in 1854. Dr. Arthur Watts of Chillicothe and Alexander Waddle of South Charleston, the man who had questioned the purity of the Sanders stock, went to Britain to make the selections. They purchased twenty cows and heifers and nine bulls. One of the bulls was the famous New Year's Day, who was bought by C. M. Clark of Clark County for $3,500. Through the 1850's he was exhibited at all the leading shows in the United States and took first prize every time. Although New Year's Day did not have many opportunities to serve cows, he sired Lady of Clark, who afterward went to Illinois, and Flora Belle, who had been bred by Robert Corwin from Scottish Bluebell. The same year that the Clark County Company was organized, the Clinton County Association imported seventeen cows and heifers and ten bulls.[65]

[63] Robert G Corwin to Brutus J. Clay, August 20, 1855, January 16, 1859, Cassius M. Clay collection.
[64] Jones, "The Beef Cattle Industry," 304; Morrow Scrapbook; *Ohio Farmer* (Cleveland), June 7, 1856.
[65] A. Sanders, *Shorthorn Cattle*, 260-61, 259.

The United States Cattle Show held at Springfield, Ohio, in the fall of 1854 brought together perhaps the biggest assemblage of cattlemen ever to have occurred in the Ohio Valley— the Steddoms, the Pierces, and the Renicks, Brutus Clay, Alexander Waddle, Arthur Watts, Solomon Meredith, and scores of others. The showyard contest between Kentucky and Ohio was on. The Kentuckians exhibited Bedford's famous Laura and Abram Renick's Rose of Sharon cow Duchess. But imported Duchess, by Norfolk, won for Ohio premier honors among the cows shown. Bedford's big, light-roan show bull Perfection (belonging to the Louan and Seventeen tribes), who had been taken on a tour of four fairs in Kentucky to test his superiority, was bested by the bull New Year's Day, who had been imported that very year by the Clark County (Ohio) Company. The Kentuckians apparently recognized New Year's Day's superiorities, for they went into a huddle with old Abe Renick at the show grounds and came up with an offer to buy the Ohio bull at a considerable advance. Reversing the precedent of several years before, when they had let their triumphant Comet Halley be sold into Kentucky, the Ohioans this time rejected the offer and kept their prize-winning bull in Ohio hands. For the sweepstakes, the herds were all lined up in a row, each headed by its bull. By the time the judges were ready to ballot for the sweepstakes, funds were so low that Brutus Clay secretly suggested to the judges a tie—and so it was, a tie between himself and Watts.[66]

Perfection, meanwhile, had taken second prize at Springfield, beating all entries except New Year's Day. The Kentuckians, observing sarcastically that the latter bull had scarcely arrived from Scotland in time for the show, claimed victory over Ohio. There ensued a vigorous but polite exchange of barbs across the Ohio River. Brutus Clay crowed that Perfection, deep in Sanders stock, had proved his superiority over the best bulls

[66] Sanders, 260-61; *Ohio Cultivator*, November 15, December 1, 1854. Waddle withdrew from the grounds his newly imported stock, which had arrived from abroad so recently that they were still somewhat seasick.

Ohioans had been able to breed—a vindication of the Sanders stock at long last! Harness Renick of Pickaway County replied that Ohio had not exhibited all its best stock; that the appetite of Ohioans for premiums had long ago been glutted. A few months later, Dr. Watts of Chillicothe broke ranks with his fellow Ohioans by declaring that the descendants of the Sanders stock were just as good as the Bates importations.[67]

The next year, the Lexington Fair pitted Kentuckians against Kentuckians. Alexander's Lord John was the only entry in the ring of aged bulls and so took both premium and certificate. In the category for other bulls, Alexander's Grand Master took the premium and John and Albert Allen's Senator the certificate, "the judges," reported one observer, "having tied and called in as umpire the same man who decided in Grandmaster's favor last year." In the third ring, for animals two years old and under, Alexander's Sirius won the premium and Bedford's Cyrus the certificate. Sirius, it will be recalled, had been imported the year before by the Scott County Importing Company and bought at auction by Alexander for $3,500. In the sweepstakes, open ring for all ages, Lord John took the premium and Sirius the certificate—both Alexander entries! Alexander went home, thus, with most of the ribbons and cash, and some of the other cattlemen went home with hard feelings.[68]

Meanwhile, Captain Brown was thinking about emulating in Illinois the example of the importing syndicates. Accordingly, the Illinois Importing Company was formed in 1857. The company purchased twenty-one cows and ten bulls, of whom one heifer and three bulls died during the stormy sea voyage. The auction was held at the Springfield fairgrounds on August 27. The company prohibited sales to non-Illinoisans, but numerous breeders came as spectators from other states. Unfortunately for the Illinoisans, the panic of 1857 struck a few weeks after the auction. Brown himself had peculiarly bad luck; the roan two-year-old heifer Rachel, for whom he

[67] *Ohio Farmer* (Cleveland), February 4, 24, 1855; H. H. Hankins to Brutus J. Clay, July 23, 1855, Cassius M. Clay collection.
[68] Johnston Farm Book.

paid $3,025 (an extraordinary price for any cow), lived but a few years and had no produce that proved fruitful. Illinoisans continued, as before this importation, to buy breeding stock from Kentuckians such as Brutus Clay.[69]

The upbreeding and importations between 1783 and 1860 had important practical results. A race of cattle was bred that could utilize the rich feed of the Ohio Valley fully and in the exact way the feeder wanted. Thus, the improved Shorthorn had early-fattening properties but more often than not was kept from the shambles until four or five years of age; at 800 to 1,500 pounds he was right for droving, but he could put on 2,000 pounds before going on the railroad cars. The beef of these big animals, though perhaps too fatty for modern tastes and sedentary people in this era of vegetable oils, was a revolutionary improvement over anything hitherto known in America. The American people developed a taste for prime beef, and the butchers demanded it at the stockyards. Accordingly, the butchers classified the droves first as common or grades or Durhams, and then sometimes further classified animals of the latter two categories as to their genealogy: out of Henry Clay's stock, or Renick's, or Selsor's.[70] The advance in breeding was a job done by hundreds of breeders, large and small, intent upon developing better fattening animals. Finally, years later, when the cattlemen of Montana began upbreeding, they looked not only to Iowa, which already had Ohio Valley cattle, for breeding stock, but also back to the Ohio Valley; for example, the bull Kirk, bred from Sharon of Highbank (the Harness-Vause farm), was taken from the Scioto Valley to Montana in 1881.[71] Seven years later the Shorthorn business collapsed, and Herefords swept to popularity—a development that occurred, some cattlemen now think, not because of the superior hardiness of the Hereford, but because "the Shorthorn breeders had been selling pedigrees instead of individuality."[72]

[69] A. Sanders, *Shorthorn Cattle*, 276-82; Brutus J. Clay, Register.
[70] See, for example, *New York Tribune*, May 3, 1855.
[71] J. I. Vause Herd Book, in possession of Joseph Vanmeter, Chillicothe, Ohio.
[72] Ben Douglas Goff, Sr., interview with the author, Winchester, Kentucky, April 11, 1957.

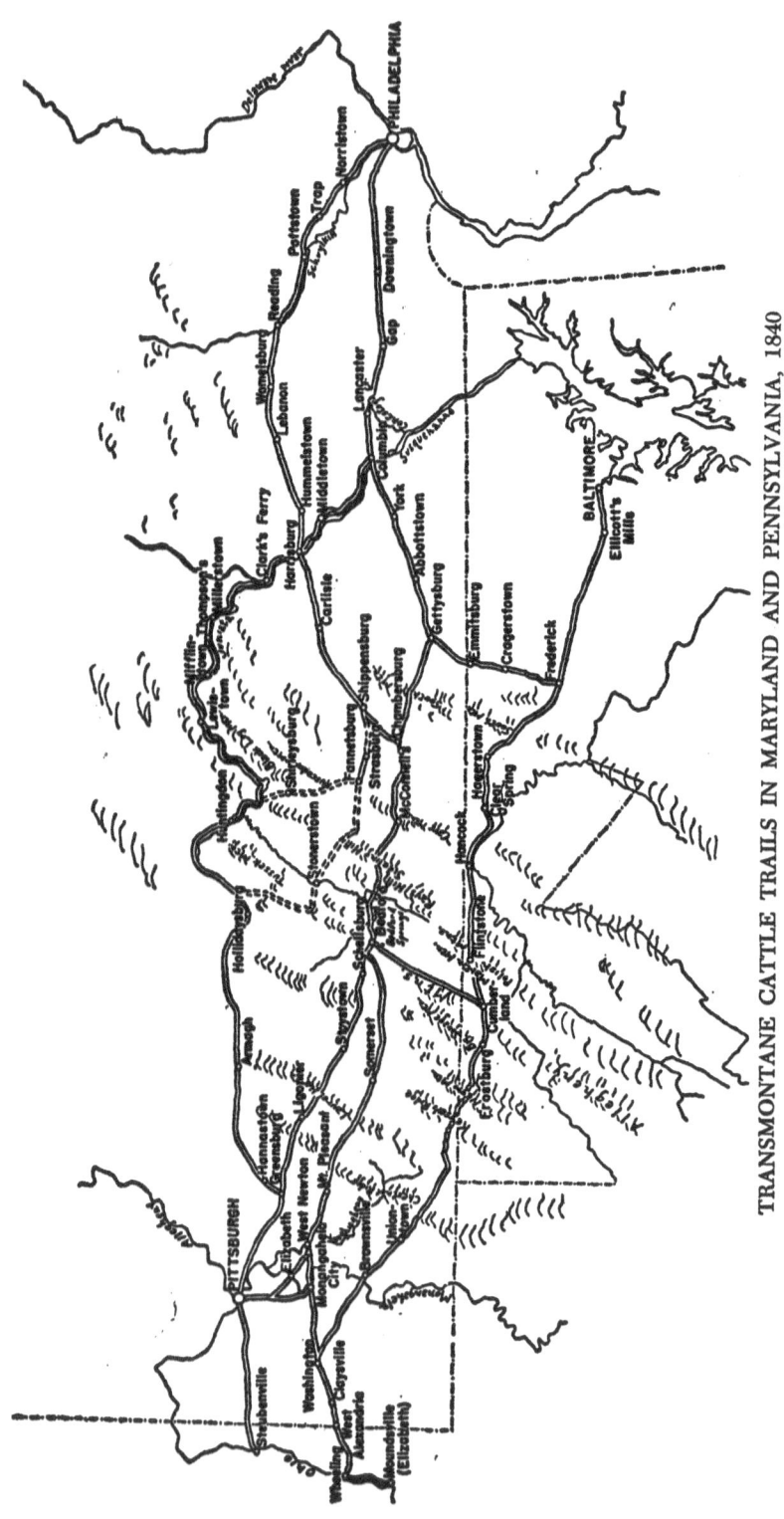

TRANSMONTANE CATTLE TRAILS IN MARYLAND AND PENNSYLVANIA, 1840

Chapter 5

The Drives over the Mountains

WHO FIRST THOUGHT OF DRIVING OHIO VALLEY CATTLE OVER THE mountains to the eastern markets is unknown, but the first man actually to drive corn-fat cattle from what is now southern Ohio was George Renick, in the spring of 1805. This was the herd of sixty-eight head, already mentioned, which he had fattened the preceding winter. He trailed them to Baltimore, that city thus becoming the first on the seaboard to see fat cattle that had been driven from the new state of Ohio. Many more long drives across the mountains followed during the next fifty years, but then, like the more famous Texas drives of a later day, the eastern drives from the Ohio country disappeared.[1]

Cattle droving began in colonial America as soon as there was an agricultural hinterland far enough removed from the towns so that it was no longer economical for a farmer to take his cattle to the town butcher one at a time. The fact that pasturage near the towns had to be saved for dairy cows and horses forced beef-cattle grazing inland. Gradually, characteristic patterns developed. In the 1790's the rougher hill districts east of the Connecticut Valley were producing cattle which were driven directly to Boston, or else to the Connecticut Valley for fattening and thence to Boston. Meanwhile, the Scotch-Irish of the Pennsylvania back country were driving their cattle into the limestone area of southeastern Pennsylvania—around York and Lebanon—where German farmers bought and stall-fed them on corn. These thrifty "Pennsylvania Dutch" feeders, with their corn and wheat fields, chicken yards, butter houses, and big sheds full of tobacco, were also taking thin cattle driven from the Carolina Piedmont or from the salt marshes along the coast. Before the Jay Treaty of 1794, thin cattle were being driven from New Jersey to the posts still retained by the British along the Canadian border. F. A. Michaux saw cattle being driven eastward through Cumberland Gap—on the "Kaintuck hawg road"—in 1802, and probably this had been going on for some years; a drive had occurred as early as 1793 from Kentucky to the Winchester area of the Shenandoah Valley. In 1800 Ephraim Cutler, a merchant and land speculator of Amesville, in the Hocking Valley, began to export eastward the cattle that he received on store bills and in exchange for land. By 1810, drovers were herding Genesee country cattle to Fort Niagara, New York City, Philadelphia, and even Baltimore. After 1814, herds were driven to New York City from Putnam County and the hilly region around it, and, as the demand increased, from the Mohawk and Cherry valleys. During the 1820's, cattle belonging to the Wadsworth family of the Genesee country became familiar sights at the Bull's Head market in New York City. By this time, cattle from Ohio were sporadically marketed at the eastern stockyards, and, indeed, some of the Wadsworths' animals were purchased in Ohio and then driven to the Genesee country for pasture fattening.[2]

As ranges and feeding areas moved west of the mountains, it became more difficult to get the cattle to the eastern butchers. It was more than eight hundred miles from the Sangamon to the Atlantic, and more than four hundred from the Scioto to the Atlantic; in between, half the stream crossings had never been bridged, and half the bridges were periodically floated away in spring floods. And then too, there was the hump of the Alleghenies to be crossed. Men said the cattle would never get through, and that if they did, they would be walking skeletons, worthless at the stockyards. But the Renicks, Jonathan Fowler,

[1] A part of this chapter appeared in *Agricultural History* (April, 1954), 28:83, and the passages therein are reproduced here by permission of the Agricultural History Society.

[2] Charles Townsend Leavitt, "The Meat and Dairy Livestock Industry, 1819-1860" (dissertation, University of Chicago, 1931), 30, 73-74; Neil Adams McNall, *An Agricultural History of the Genesee Valley, 1790-1860* (Philadelphia, 1925), 94-95; F. A. Michaux, "Travels to the West of the Allegheny Mountains," in Reuben Gold Thwaites (ed.), *Early Western Travels* (Cleveland, 1904-1907), 3:245; Gilbert Imlay, *A Topographical Description of the Western Territory of North America* (3d ed., London, 1797), 61, cited in Robert Leslie Jones, "The Beef Cattle Industry in Ohio prior to the Civil War," *Ohio Historical Quarterly* (April, 1955), 64:173; Bouck White, *The Book of Daniel Drew* (New York, 1911), 81-82.

and Thomas Goff, pointing to the drives on the Wilderness Road, disagreed.

By the 1820's, herds of cattle were plodding east along six busy mountain trails (listed from south to north): the Wilderness Road, the Great Kanawha route, the Cumberland Road, the Pittsburgh-Chambersburg route, the Greensburg-Juniata road, and the Erie-Mohawk trail. Probably the first transmontane cattle to reach the eastern market walked over the Wilderness Road ("Kaintuck Road") through Cumberland Gap, Kentucky's historic connection with the East, well before 1800. Livestock worth $1,167,000 passed through the tollhouse at the Gap in 1828.[3] This route proved more popular for cattle and hogs than for sheep. It was well located to serve the Bluegrass; but the farmers of Gallatin and Carroll counties, located to the north on the lower Kentucky River, would hardly choose to send cattle east this way. The fastest route from Gallatin County to Baltimore was by way of Cincinnati and the Miami Valley pikes which led to the National Road west of Columbus.

To the north of the Wilderness Road lay a second cattle trail, beginning at the bloodied ground of Point Pleasant, opposite Gallipolis, Ohio, and running up the Great Kanawha to Charleston, where it joined (1831) the James River and Kanawha turnpike. Cattle, hogs, horses, men, wagons, and coaches jostled in the streets of Charleston. Farmers in the Kanawha bottoms below Charleston, and along the turnpike, sold much of their corn to drovers—a reason why this part of the upper Ohio Valley never developed much of a cattle-feeding industry of its own. One authority says this route was important for swine from both Kentucky and Ohio, but not for cattle; local tradition in Clark County, Kentucky, however, indicates that the Great Kanawha route was repeatedly used in driving cattle to the Baltimore market, a favorite with the cattlemen.[4] Drovers especially liked the plentiful sources of salt for their

[3] Frederick J. Turner, *The Rise of the New West* (New York, 1906), 102.
[4] Jones, "The Beef Cattle Industry," 294; Ben Douglas Goff, Jr., interview with the author, Winchester, Kentucky, October 25, 1955.

animals along the way, but the route, like the Wilderness Road, was of little value for reaching the Philadelphia or New York markets. Even the Baltimore market was somewhat more directly accessible to southeastern Ohio by the Cumberland Road than by this route.

The Northwestern Turnpike, lying just to the north of the Great Kanawha route, ran from Parkersburg across the mountains to Winchester (approximately U.S. 50). After the opening of the Cumberland Road in the 1820's, however, this route was no longer a major drove road.[5]

The third route, the Cumberland Road or Pike, later the National Road, was superbly located to tap the various trans-Allegheny cattle regions. It is true that the Illinois portion of the road ran well south of the cattle regions of that state and that there was no bridge on the National Road at Terre Haute; yet the route was much used in droving from Illinois because there was no other usable road at all crossing the Illinois-Indiana border for nearly one hundred miles north of Terre Haute. At Indianapolis the Illinois cattle met herds of Indiana cattle, for into the town from the northwest came an excellent road network, partly macadamized by 1840, which served the Indianapolis–Wabash Valley fat-cattle district. From Indianapolis the National Road ran almost straight east, across the upper Great Miami, to Columbus, Ohio. Cattle bound for Ross County for feeding sometimes turned off the road at Springfield, Ohio. An excellent system of roads led up the Miami Valley to the National Road; the South Charleston cattle district was almost on the great thoroughfare; and none of the cattle counties of Ross, Pickaway, Licking, or Fairfield were far from it. Cattle from Licking County would strike the National Road east of Columbus, at or near Jacksonville, and droves from Fairfield County would track onto the National Road at Zanesville, where by 1840 a covered wooden "Y" bridge spanned the confluence of the Licking and Muskingum rivers. These and other cattle from eastern Ohio, and most of the Kentucky

[5] Jones, "The Beef Cattle Industry," 294.

cattle, comprised the only important eastbound herds which did not funnel through the Scioto river towns of Columbus or Chillicothe. So great was the volume of cattle crossing the Scioto for the East that the great eastern cattle drive may be called the Scioto drive.

The pioneer of cattle droving on the Cumberland Road was, so far as we know, George Renick, who in 1805 drove cattle from the Scioto Valley to Wheeling, by Zane's Trace, and thence through Cumberland and Frederick to Baltimore, by the Cumberland Road. As the Cumberland Road was extended westward, other men followed Renick's example, and soon this great highroad of America—the embryo United States Highway 40— teemed with dusty herds going east.

More of the streams on the National Road were bridged than on other roads. The Brownsville covered wooden bridge, the first bridging of the Monongahela upstream from Pittsburgh, dated from 1833. In fact, the National Road even had a bridge over the Ohio at Wheeling by 1840, the first bridging of that great river.

Eastbound traffic on the National Road faced an important fork at Cambridge, Ohio, where the great thoroughfare itself wound east to Wheeling. The road branching off, taken by most of the drovers bound for New York and Philadelphia, ran northeastward to Steubenville and Pittsburgh. Cattle bound for Baltimore stayed on the National Road. Their drovers could still reach Philadelphia by turning off at Cumberland or still later at Frederick. Whichever fork the drover took at Cambridge, the going was tough in eastern Ohio in the early days. It was said in 1820 that emigrants from the East "on the Wheeling and Steubenville roads ... encounter hills more steep and difficult than the mountains they have passed for near 200 miles to the vicinity of Chillicothe many curse the road and country: break their wagons, wear down their horses to the bone."[6]

Most of the cattle passing through Pittsburgh came from

[6] Richard T. Wiley, *Monongahela, The River and Its Region* (Butler, Pennsylvania, 1937), 53-56; *Western Star* (Lebanon, Ohio), February 22, 1820.

Steubenville; but those few coming from Medina County were just as likely to come by way of Wellsville, where there were salt springs and, as at Steubenville, a ferry across the Ohio. Presently there was enough population at Pittsburgh, and enough cattle arriving there, for a local cattle market to arise. The slaughterhouses in Pittsburgh's Bayardstown were in operation by the 1840's.

The fourth and fifth major cattle trails were those running east from Pittsburgh to Philadelphia, one by way of Chambersburg, the other by way of Greensburg and the Juniata River. The Chambersburg road had several alternate routings at its western end and also several at its eastern end. No matter which of the alternate ways to Chambersburg the eastbound drover chose, he had to climb the Laurel Hill and a saddle in the Sideling Hill several miles south of the site of the present Pennsylvania Turnpike Sideling Hill Tunnel. The road between Pittsburgh and Chambersburg via Ligonier was known as Forbes Trace and has been followed approximately by United States Highway 30, the Lincoln Highway. The alternate routing between Pittsburgh and Bedford via West Newton and Somerset was called the Old Glade Road, and, between West Newton and Somerset, is approximately the route of Pennsylvania state highway 31. The three alternate routings at the eastern end of the Chambersburg road were the Carlisle-Reading, the Carlisle-Columbia, and the Gettysburg-Columbia. The old trail between Chambersburg and Carlisle is almost exactly the roadbed of U. S. 11; this trail had the advantage of avoiding the South Mountain, known to geographers as the "Cumberland Prong," but had nothing to compare in beauty with the descent of South Mountain through the apple-orchard slopes into Gettysburg. The Cumberland Road and Forbes Road were connected by the Cumberland-Bedford road, which ran past Bedford Springs, and the Frederick-Gettysburg road, which ran through the German settlements along the Monocacy. Thus, drovers on their way to Baltimore might switch to the Philadelphia market.

The Greensburg-Juniata route, except in the Greensburg-Hannastown neighborhood approximately the route of U. S. Highway 22, utilized the natural passage through the mountains provided by the Juniata River Valley, but it was less used by drovers than the Chambersburg route, partly because of the apparently greater distance. A contemporary gazeteer estimated it to be only five miles farther from Pittsburgh to Harrisburg via the Juniata than via Chambersburg. A more important reason probably was the sparseness of agricultural population and the consequent difficulty in getting feed for the drove, although around McVeytown (75 miles from Harrisburg) there were some good stock farms.[7]

The sixth cattle trail ran through Erie (Pennsylvania), Dunkirk (New York), Geneva, and Fort Schuyler. The stretch between Geneva and Fort Schuyler comprised the Great Genesee Road or the Ontario and Genesee Turnpike, 100 miles long, four rods wide, partly gravel or log corduroy.[8] Genesee Street in Utica and Genesee Street in Syracuse were both once clogged with cattle walking from the Genesee Valley or from Ohio on their way to market over this turnpike. Since 1800 (perhaps earlier) there had been a bridge over the Mohawk at Fort Schuyler, at the foot of Genesee Street. When the Erie Canal was built, it took some of the cattle traffic once using the parallel land route. Droves from central Ohio could reach this route via Mount Vernon, Wooster, and Cleveland, or via Zanesville, Akron, and Ashtabula; if they were bound for Boston, they would probably take this Erie–Fort Schuyler route, and if they were bound for New York City, they might take it.

Most herds from the Ohio country were started on the trail any time from the middle of February until the first or middle of June, and reached the eastern stockyards between April 15 and August 15. By the middle of August, western cattle had to

[7] John Melish, *A Description of the Roads of the United States* (Philadelphia, 1814), 55; J. Gould, "Wanderings in the West in 1839," *Indiana Magazine of History* (March, 1934), 30:75.
[8] Albert Perry Brigham, "The Great Roads across the Appalachians," American Geographic Society of New York, *Bulletin* (1905), 37:326.

compete in the New York market with the grass-fed herds of New York state. Hence only a limited number of grass-fed and some corn-fattened cattle were sent from the West to arrive at the market in December. The Philadelphia market had its greatest receipts of beeves in June, but receipts were heavy in December when fall-driven western cattle were coming in. Spring-driving meant quagmire roads and swollen streams to contend with; by June the rivers were down, but the herds on the rutted trails raised long streamers of dust which could be seen for miles. Early spring droving starting in February, or late fall droving, was likely to encounter frozen roads, which were thought to be wearing on the animals, perhaps even laming them.[9]

By 1840 the drives also fed the western markets, chiefly Chicago and St. Louis. There were two categories of drives: those of thin cattle from range to feeding area, and those of fat cattle from feeding area to market. Cattle from the trans-Mississippi range after 1846 might go to Chicago, or to the Sangamon feeders, or to the Sangamon via the Chicago market. After about 1815 Illinois thin cattle began to be driven eastward on the National Road, most of them to be fattened in the Miami or Scioto valleys; fat cattle from the middle Wabash and the Miami valleys hit the National Road at Indianapolis and Springfield, Ohio, respectively; Champaign County cattle struck the road at or near Summerford or at Columbus; the Scioto Valley disgorged its fat cattle onto the road at Columbus, Jacksonville, or Zanesville; the number passing through Zanesville was increased by Licking and Fairfield county cattle. Some three-year-olds from the west were purchased by local feeders as far east as Lebanon, Pennsylvania, there to be fattened and sold in the nearby Philadelphia market.[10]

[9] *Western Farmer and Gardener* (Indianapolis), November 15, 1846; *Pennsylvania Inquirer* (Philadelphia), December 31, 1847; *Dollar Farmer* (Louisville), March, 1843.

[10] Stevenson Whitcomb Fletcher, *Pennsylvania Agriculture and Country Life, 1640-1840* (Harrisburg, 1950), 180; J. F. King, "The Coming and Going of Ohio Droving," *Ohio Archaeological and Historical Publications* (April, 1908), 17:247.

It was believed that purebred Durhams, because of their heavier weight, did not travel on frozen ground as well as the common animals. Cattle driven from the McLean-Champaign county range of Illinois to the Scioto Valley of Ohio and fattened there were thought by some to take the long drive better than animals of Ohio origin.[11] Drovers encountered mud and dust on the dirt roads, but on the macadam roads sharp angular stones cut the hooves of the cattle. One cattleman warned, "a paved road will ruin [a bull] in going 10 miles if forced." Many drovers found it necessary to shoe their animals. Blacksmiths along some macadam roads had machinery to lift the heavy steers off their feet so that they could be shod.[12]

The weight lost by an animal on the drive was known as "drift" or "shrink." Both sheep and cattle, as a rule, suffered less drift than did hogs. But the drift depended to a large extent on the amount of feed provided for the animals on the way; one Kentucky drover reported that his hogs suffered no drift. Indeed, they averaged 150 pounds at starting and 180 pounds upon arriving at New York, a gain of nearly one-half pound per day on the trail. The usual weight at which to start a fat steer on the transmontane trail was 900 to 1,000 pounds. Steers heavier than 1,000 pounds apparently suffered a greater percentage of flesh loss on the drive. Nevertheless, local tradition in Clark County, Kentucky, and some other evidence as well, indicates the cattle were driven at 1,200 to 1,500 pounds. Heavy cattle suffered disproportionate drift in hot weather. A bullock starting at 1,000 pounds as a rule lost at least 150 pounds; one of 900 pounds lost from 100 to 150 pounds.[13] The

---

[11] *Dollar Farmer*, March, 1843; William Renick, "On the Cattle Trade of the Scioto Valley," in Ohio State Board of Agriculture, *Third Report* . . . 1848, p. 163.

[12] Nathaniel Hart to Virginia Shelby, April 24, 1841, in possession of the Filson Club, Louisville, Kentucky; King, "Ohio Droving," 251.

[13] *Western Farmer and Gardener*, November 15, 1846; William Renick, *Memoirs, Correspondence, and Reminiscences* (Circleville, Ohio, 1880), 28; Ben Douglas Goff, Sr., interview with the author, Winchester, Kentucky, November 9, 1955; Charles Augustus Murray, *Travels in North America during the Years 1834, 1835, and 1836* (London, 1839), 224. Tom Shelby's and John Webb's cattle apparently weighed an average of slightly less than 800 pounds each upon arrival at the Bull's Head in New York in 1843. Thomas H. Shelby to Isaac Shelby, May 14, 1843, in University of Kentucky Library.

distance to be traveled also had, of course, some affect on the drift.

There were four types of drover: the cattleman who was his own drover; the hired man who did the driving; the agent who drove and kept informed on market conditions; and the freelance professional drover. This last type is one of the picturesque figures of American history. The consumer resented him as the middleman who kept the price of beefsteak high. The farmers sometimes saw him as a cheat. But some recognized his value: "We should be arrant blockheads to leave our crops ungathered, our fields uncultivated, our homes and our farms neglected, for the sake of ministering to the cupidity of those [the butchers] who are laboring to reduce the price of our stock."[14]

The professional drover, a figure known from Boston to the Indian country, mounted on his horse, with saddlebags and snake whip, would ride into a community and spend two months during the winter buying up cattle for the spring drive. Usually he earned a reputation for honesty in his dealings with the farmer. One such was the likable Benjamin Franklin (Frank) Harris, who drove cattle for eight years (1834-1842), mostly from Illinois, before ever raising any himself or even owning a farm.[15]

An unscrupulous drover, however, might try to cheat the farmer by taking his cattle on credit. After he had acquired an evil reputation in one neighborhood, he would shift his operations into another. A really shifty drover would do a lot of haggling with the farmer about the price of the steer, so that the farmer would be deluded into figuring, "A fellow who drives such a sharp bargain must surely be as good as his word."

[14] *Monthly Genesee Farmer* (Rochester, New York), July, 1839.
[15] McNall, *An Agricultural History*, 95; Mary V. Harris (ed.), "The Autobiography of Benjamin Franklin Harris," *Transactions of the Illinois State Historical Society for the Year 1923*, p. 72. One of the Renicks in the winter of 1853-1854 bought 1,200 head of cattle in northern Texas. W. Renick, *Memoirs*, 24. On the basis of the origins of the Texas population, we may assume that many of the early Texas cattlemen practiced droving in Kentucky, Ohio, and Illinois.

In 1842 some such unscrupulous professional drovers were going around Ohio buying cattle with the worthless paper of bankrupt eastern banks.[16]

The Wadsworth family of the Genesee country, using hired men and agents as drovers on a large scale, sent their men into Ohio to purchase yearlings and two-year-olds and drive them eastward. This was usually done in the spring so that the cattle could feed on grass on the trip. The agents were given money to pay cash for the cattle and to hire men for the drive. Once the cattle reached the Genesee country, they were pastured there for a while before being marketed.[17]

Strauder Goff of Clark County, Kentucky, marketed many of his cattle through an agent-drover, usually B. F. Cloud. By 1848 Goff had worked out a system by which James Gay was in Charleston, South Carolina, getting contracts from the butchers for Goff's cattle and Cloud was doing the actual droving or else directing it. How this system broke down will be told later.[18]

One type of agent-drover was the butcher's agent, who appeared frequently in Kentucky. The Baltimore butchers regularly sent agents to Clark County to assemble herds. The herds were held at Winchester until the agents learned the sailing dates of ships from Baltimore for the English market. In 1843 one of the slaughterhouses, Vanbrant and Adams, dispatched J. C. Hughes to the Ohio Valley to contract for cattle. Strauder Goff agreed to deliver to Hughes in Cincinnati between November 10 and 20 one hundred twenty head of fat beef cattle, for which Hughes was to pay three dollars per hundred pounds upon delivery.[19]

A favorite place for professional drovers to pick up cattle was the tricounty area around South Charleston, Ohio.[20] Here fat cornfed beeves and range cattle were to be had. Ranking second only to Chillicothe as the best known cattle town of

[16] White, *Daniel Drew*, 28-32; *Scioto Gazette* (Chillicothe, Ohio), June 23, 1842.   [17] McNall, *An Agricultural History*, 135.
[18] James Gay to Strauder Goff, April 28, 1848, in possession of Ben Douglas Goff, Sr., Winchester, Kentucky.
[19] Strauder Goff to J. C. Hughes, August 28, 1843, Ben Douglas Goff, Sr., collection.   [20] *Ohio Cultivator* (Columbus), August 1, 1845.

Ohio, South Charleston sent its herds onto the nearby National Road.

In Illinois, where some ranches in McLean and Champaign counties covered thousands of acres, the professional drover would appoint a time and place where all the cattle he had purchased in the vicinity were to be assembled. An eyewitness has given us a description, perhaps more dramatic than true, of what happened at the beginning of one drive. It was sunrise, and the horses stood tied to young hickory trees, while the men with heavy whips in their hands waited for the signal to start. The cattle rushed from the pens, "pell mell amidst a torrent of shouts and yells. The charge was desperate and for some time it was hard to say whether the beasts or their opponents would gain the victory."[21]

The numbers of cattle fed in the Scioto Valley and the Kentucky Bluegrass and the counts made of cattle actually passing along a transmontane trail are evidence that there must have been hundreds of drives between 1800 and 1850, of which we have record of only a few. Ephraim Cutler says he drove a herd from Marietta to the headwaters of the Potomac in 1800. Captain Jonathan Fowler of Poland, Ohio, drove a large herd which he had purchased in the Canfield area (in the Western Reserve) to the Philadelphia market in 1804.[22] After George Renick's famous drive to Baltimore, Cutler was driving thin cattle to York, Pennsylvania, and Allegheny County, Maryland, in the years 1809-1812. George Renick's brother Felix, best remembered for his work in the Ohio Company for Importing English Cattle, made his first recorded drive in 1815. While Felix was on this drive, Lewis Heath of Paris, Kentucky, was acting as his agent in Kentucky, buying thin cattle there for him. Also in 1815, a drive occurred from the Mad River of west-central Ohio, then on the frontier, to Philadelphia. In 1816, Thomas Goff of Clark County, Kentucky, drove 130 Patton cattle to the same place. On the road between Zanesville

[21] William Oliver, *Eight Months in Illinois* (Chicago, 1924), 106.
[22] William T. Utter, *The Pioneer State, 1803-1825*, in Carl Wittke (ed.), *The History of the State of Ohio* (Columbus, 1941-1944), 2:157.

and Lancaster, on June 24, 1816, a traveler encountered 300 head of cattle on their way to Baltimore or Philadelphia. The next year, a man named Drenning herded some 200 head from Chillicothe to New York. In 1818 or 1819, Daniel Drew of New York, a professional drover, took a big herd of about 2,000 across the mountains, but lost 400 to 500 during the drive. Drew's biographer mistakenly claimed this was the first drove of cattle to be driven east across the Alleghenies. Before 1818 the Wadsworths of the Genesee country had sought an agent to go to Kentucky and Indiana to purchase from 100 to 200 head of cattle; and in 1818 the family planned shipment of eight and a half tons of cheese via Olean and the Allegheny River to Kentucky, "there to vend it and to place the proceeds in a drove of young cattle to this place, keep them on the flats for one or two years and send them to the New York market." In 1818, George Renick and Joseph Harness sent a drove to New York City; William Renick mistakenly claims that this was the first transmontane drive to that place. More than one large herd of cattle and hogs were driven, too, from Pickaway County, Ohio, during 1818 and 1819.[23]

By 1820 the long drives were becoming almost commonplace. Droves probably numbering up to 1,200 head each, presumably from Kentucky, were passing between Winchester and Staunton, Virginia, about 1820 or 1821; a "Major H." near Middletown made a practice of buying some of these animals for feeding. Through Cumberland Gap in the fall of 1821 were herded 410 cattle, on their way out of Kentucky, and next year 236 stall-fed steers passed through Cumberland Ford enroute

---

[23] Felix Renick Account Book, in possession of Renick Cunningham, Chillicothe, Ohio. Published works have heretofore placed the date of Felix's first known drive to Philadelphia at 1817. James Flint, "Letters from America," in Thwaites, *Early Western Travels*, 9:80; Lucien Beckner, "Kentucky's Glamorous Shorthorn Age," *Filson Club Historical Quarterly* (January, 1952), 26:32; Charles S. Plumb, "Felix Renick, Pioneer," *Ohio Archaeological and Historical Publications* (January, 1924), 32:21; David Thomas, *Travels through the Western Country in the Summer of 1816* (Auburn, New York, 1819), 90; James W. Thompson, *A History of Livestock Raising in the United States, 1607-1860* (Washington, 1942), 94; White, *Daniel Drew*, 81; McNall, *An Agricultural History*, 95; Renick, *Memoirs*, 97; *Springfield Farmer* (Springfield, Ohio), February 13, 1819.

to the Gap. Samuel Lutz of Pickaway County, Ohio, in 1822 drove a herd of fat cattle to the Baltimore market. A drove of cattle was taken to New York City in 1824 by Richard Seymour, also of Pickaway County. Another lot was taken, three weeks later, by William Renick himself. A writer from the central part of New York state complained in 1826, "our western neighbors, who drive cattle from a vast distance to the Atlantic markets, are too powerful for us in their rivalship," but it not not entirely clear whether the reference is to Ohio or to Genesee stockmen. Young Thomas Atkinson of northwestern Indiana one fall in the 1820's drove his brother's herd of cattle to Pennsylvania, presumably to the feeders of southeastern Pennsylvania. Some 1,525 stall-fed steers were herded through Cumberland Ford in 1828.[24]

The 1830's and 1840's witnessed the greatest number of drives. In 1833, George Renick led a drive to New York City via Wheeling. The next year Frank Harris started on his first drive, from Clark County, Ohio, to Lancaster County, Pennsylvania, acting as a hired drover for Jim Foley. In November, 1834, he began his first drive as a professional drover, herding over the snow-covered mountains 100 cattle purchased from Foley. Hardly back from this drive, he assembled a herd in Morgan County, Illinois, late in June, grazed them for a few weeks, and then drove them to the feeding area of Pennsylvania. A Kentuckian living on the road to South Carolina counted 2,485 head of "stalled" beef cattle walking by on their way to South Carolina during that year of 1835. Abner Cunningham sold five lots of Kentucky cattle in New York City in July, 1836, and a man named Finch sold three lots at Philadelphia; these eight lots consisted of about 900 cattle. In the fall of 1836 or 1837, Walter Angus Dunn's brother George, who lived in Philadelphia, sold some cattle there for him. Probably also in the 1830's, Isaac Funk, newly moved from Madison

[24] *American Farmer* (Baltimore), June 29, 1821; *Western Star*, March 9, 1822; *Niles' Weekly Register*, December 28, 1822; February 14, 1829; King, "Ohio Droving," 247; Sherman N. Geary, "The Cattle Industry of Benton County," *Indiana Magazine of History* (March, 1925), 21:28.

County, Ohio, to McLean County, Illinois, sent at least one herd to Buffalo, New York. In the latter part of the decade, when John T. Alexander was between thirteen and twenty years old, this young man, who was destined to move to Illinois and become a cattleman in his own right, helped his father drive cattle from Ohio over the Alleghenies to the eastern markets. During the 1840 season 4,113 cattle were tallied passing through the turnpike gate at Asheville, North Carolina, from the West. Some 5,000 head of cattle, presumably stockers, were driven from the Peoria area to Ohio in the early fall of 1841. William Renick in 1842 took a drove of his father's Durhams to the Brighton market near Boston. General Isaac Shelby of Fayette County, Kentucky, drove a herd of Shorthorns to the Bull's Head in New York City, arriving there in December, 1842, or January, 1843. John Webb, who did some droving for the Shelby family, drove 104 Kentucky cattle into Newark on Friday night, May 12, 1843, seventy-five days from the Bluegrass; he sent nine head by rail from Harrisburg to Philadelphia. In the late fall of 1843, J. C. Hughes, agent of a Baltimore slaughterhouse, made a drive from Cincinnati to Baltimore. Also by 1843, stock cattle were being driven from the Grand Prairie of Indiana to the feeding area of Pennsylvania. Henry Clay contributed eighty fat cattle to a drove setting out in February or March, 1844, for the East. Webb took a drove of Tom Shelby's cattle to New York that same spring; rain in the mountains made the cattle tender footed and the market was already glutted with Ohio cattle. The next spring he took young Tom and went by way of Fannetsburg, which they reached on April 10, "within 8 miles of being over the mountains." B. F. Cloud, agent drover, began a drive in the late spring of 1845 "to the East," probably Baltimore. The next year John T. Alexander, then twenty-five years old, drove 205 fat cattle from Illinois to Albany, New York. There he apparently sold them to a buyer from Boston, and drove them on to Boston himself. John Webb made another drive for Tom Shelby in 1847, starting out in late February and proceeding by way of Bedford, Pennsylvania. During at least part of the

1848 season, a drove of Strauder Goff's cattle started out every ten or twelve days from Clark County, Kentucky, for Charleston, South Carolina. During these years, David Selsor was driving east regularly from the "buck prairie" of Madison County, Ohio. In 1849 someone in Morristown, New Jersey, went to the trouble of asking each drover passing through on the road to New York how big his herd was and where the cattle were from; the tally, presumably made from March 1 to August 1 and comprising most of the cattle driven over the Pennsylvania trails to New York City before the fall season, mounted to 10,983 head.[25]

Long-distance droving continued into the 1850's. From Benton County, Indiana, Charles Sumner was sending cattle on foot to New York City in the early 1850's at the rate of 100 cattle every two weeks during the growing season. A herd of 100 of Frank Harris' cattle was taken by a hired man in 1854 to Boston, part of the way perhaps by railroad.[26]

In the published accounts, and in faded manuscripts packed away in the farmhouses of the Scioto Valley and the Kentucky Bluegrass, we catch a glimpse of what the drive itself was like.[27]

[25] Harris, "Autobiography," 75; James Hall, *Statistics of the West at the Close of the Year 1835* (Cincinnati, 1836), 257; Abner Cunningham to Brutus J. Clay, July 31, 1836, in possession of Cassius M. Clay, Paris, Kentucky; George W. Dunn to Walter Angus Dunn, January 3, 1838, in University of Kentucky Library; Albert Robinson Greene, "A Little McClean County History," Illinois State Historical Society, *Journal* (October, 1915), 8:461; Clarence P. McClelland, "Jacob Strawn and John T. Alexander, Central Illinois Stockmen," Illinois State Historical Society, *Journal* (June, 1941), 34:201; *Franklin Farmer* (Frankfort, Kentucky), March 14, 1840; W. Renick, *Memoirs*, 58; U. S. Commissioner of Patents, *Report, 1843*, pp. 386-87; *Scioto Gazette*, April 11, 1844; Thomas H. Shelby to M. C. Shelby, May 4, 1844; Thomas H. Shelby to Isaac P. Shelby, May 14, 1843, April 10, 1845, in University of Kentucky Library; Strauder Goff to J. C. Hughes, August 28, 1843; B. F. Cloud to Strauder Goff, May 13, 1845; James Gay to Strauder Goff, April 28, 1848, in possession of Ben Douglas Goff, Sr., Winchester, Kentucky; David Selsor Graham, interview with the author, Midway, Ohio, September 9, 1955; *Ohio Cultivator*, October 15, 1849.

[26] Paul W. Gates, "Hoosier Cattle Kings," *Indiana Magazine of History* (March, 1948), 44:15; Harris, "Autobiography," 88.

[27] Fletcher, *Pennsylvania Agriculture*, 180; Edward Norris Wentworth, *America's Sheep Trails* (Ames, Iowa, 1948), 67; *Western Farmer and Gardener*, November 15, 1846; White, *Daniel Drew*, 84-85; King, "Ohio Droving," 249; and the Renick papers in possession of Renick Cunningham.

The herd or drove seldom numbered less than 100 head, usually about 100 or 120 head, sometimes up to 200 head, occasionally a thousand or more. If the cattle were fat, a drove of thin or stock hogs might go along to eat the droppings of the cattle and the grain that the cattle wasted. A drove of horses could travel twenty-two miles per day; a drove of stock cattle, nine miles; fat cattle, seven miles; cattle with hogs following, five miles. The herds moved more slowly across the mountain ridges such as Keyser's Ridge or the Polish Mountains on the National Road and Laurel Hill or Sideling Hill on the Pittsburgh-Chambersburg road, but moved faster on easy sections of the road.

The crew consisted of a "boss" and one to five others. The boss rode on horseback, with saddlebags which contained a change of "linen" for himself and the men who were afoot; on his saddle was a roll of extra garments for use by the crew in stormy weather. He was armed with a good "blacksnake" or "Centerville" whip. In the case of large droves, a second man might also be mounted. Many drovers took no crew but depended on picking up "pike boys" along the route to help them with the herding. Sometimes the boys could go for one day only and had to return to their parents at night. In other instances they were allowed to be gone for two days, and, through the Pennsylvania mountains, where farms were farther apart than in Ohio, they were usually employed for several days. Although there was plenty of itinerant labor in the Miami Valley and perhaps in the Kentucky Bluegrass, the cattleman had a relatively limited choice of men known to be trustworthy enough to be given charge of the drove. William P. Hart recounted his difficulties in finding an assistant: "Hudson declined going—Nicoff disappointed me when everything was ready for a start, horses saddled and hands waiting. I had to send to Mercer [County] to know what had become of him—he was in Washington County waiting on a sick uncle having entirely declined going—Boice came over, I employed him, four days after he wrote me he could not go—I went over to Anderson to get Tom Elliston, he came over but declined going.

Then to Fayette [County] and employed young Downing he took sick and could not start—At last I started them, Eugene Cooke . . . as Purser, Young Edwards . . . as a hand."[28]

Sometimes the professional drover wanted a man who could take charge in his absence. The journal of a young Delawarean tells how he was hired for this purpose: "the Dr. Said Mr. Stite [the drover] can not be with the Drove all the time, he has bills to collect for horses sold last Spring on other Roads want a good man to take charge of the Drove pay toll make arrangement at Hotel for the drove and men, pay all bills in his absence." Then the drover and his prospective assistant bargained about the wage: "I told Mr. Stite I would like to go as his assistant. He Said he would give fifteen dollars did not want me to work but see the men attend well to the Drove and pay all bills in my absence. I do not think that enough I will have a good deal of care and responsibility. I will do the very best for you I can. I will give you 20$ I will go for 25$ Dr. Ham then said Sitte be liberal I know this young man will Suit you first Rate. well he Said I will give 25$ as the Dr. thinks I will be well Suited."[29]

The order of march varied somewhat. Usually, one man or boy on foot went ahead, guiding the "lead" ox with a rope tied to its horns; the boss, on horseback, rode behind the first forty head; a third man, usually afoot, brought up the rear, applying his "Centerville" to lagging steers. If the entire crew was mounted on horseback, one of them usually had the job of riding ahead and jumping off his horse to pay tolls; the tollgate keeper would usually take his word for how many cattle were in the drove.[30] The boss was often not to be found with the herd at all; he would be riding ahead, looking for a place for the outfit to "noon" or for a good place to cross the next

[28] William P. Hart to Virginia Shelby, August 27, 1841, Filson Club.
[29] Marie Windell (ed.), "James Van Dyke Moore's Trip to the West, 1826-1828," *Delaware History* (September, 1950), 4:102.
[30] Thomas M. Donovan to Thomas H. Shelby, March 19, 1847, in University of Kentucky Library. The temptation to cheat the turnpike company was too much for some drovers. One drover—and a hired one at that—who declared 110 cattle at Licking Bridge was observed to have 120 upon arrival at Maysville.

river, or he might be making arrangements at a "drove stand" for spending the night; sometimes he would be far away on another road, collecting money for animals sold along the way on last spring's drive, or paying farmers for animals taken on credit during the last drive. Just as the humans had an order of march, so the animals drifted into one; in the case of a mixed drove, the sheep after the first day or two got to know the cattle and crowded in close behind them without any urging; the cattle marched, seriously and soldierlike, abreast of one another, the animals behind walking in the footsteps of the ones ahead, and thus hoofing the roads into what swearing stagecoach drivers called "cattle billows."

As the dawn broke, the boss would be up and around, looking over the herd to see if any animals had died during the night. Next he would go up to the keeper of the "drove stand," which was sometimes an inn but usually just a large farmhouse, to pay the night's bill, according to the rates previously agreed upon. Overnight grazing was usually five cents a head, though sometimes it would be obtained for less—perhaps a dollar for the whole herd of 100; forage might cost ten dollars a night for the whole herd. The corn might be purchased for around 25 cents a bushel or 37½ cents a shock; lodging ran from 37 to 87 cents for one man, to $3.00 to $9.00 dollars for the whole crew, including meals. There was not always an outright cash payment; the lodging and fodder bill could be paid by leaving with the farmer the calves that had been dropped during the night or the previous day's march, or by leaving sick animals. Other times, however, the drover paid the bill in cash and swapped the new lambs or calves for some of the farmer's fatlings.[31]

After these financial arrangements had been made, the boss returned to his herd and the day's drive got under way. Excitement was likely to come quickly. A balky steer might lie down

[31] White, *Daniel Drew*, 84-85; Renick Cunningham, interview with the author, Chillicothe, Ohio, April 22, 1953; Felix Renick Account Book; King, "Ohio Droving," 250; Wentworth, *Sheep Trails*, 67; Windell, "Moore's Trip," 102-103.

in the middle of the road and refuse to move; a recommended remedy (which may have resulted in some gored drovers) was to clasp the hand over the beast's nostrils until he stirred and started up. Some animals might stray, or the herd might get tangled in other traffic. If it were Sunday morning, church bells might set off a stampede. At midday the whole outfit halted in some well-shaded lot to "noon"; while the cattle were resting and feeding, the boss would scout a river to be crossed in the afternoon. If there was no bridge and the water deep, the river crossing was likely to be the crisis of the day. One man might be sent across to tie a rope to a tree on the opposite bank and the cattle forced to swim across, using the rope as a guide. If there was a ferry, the lead ox and some ten bullocks might be put on board to serve as leaders and the rest of the herd driven into the river to swim in the wake of the ferry. Summer drives, however, might find the streams dry. In the drought of 1839, the Ohio River fell so low that on August 21 the channel was only twenty-one inches deep at Pittsburgh, a focal point of cattle trails.[32]

If the outfit came to a drover inn, such as "Aunt" Hetty Jackson's on the National Road east of Wheeling, during the afternoon, the men might stop for cider or "beer and cakes." At night the outfit would put up at the stand selected in advance by the boss, and the men would grumble if the food and beds were not as good as at Ezekiel Bundy's famous stand near Barnesville, Ohio.[33] After supper, the boss would check up on his animals, looking for signs of injury or sickness. If all were well and the animals were behind fences and contentedly feeding, he might relax, perhaps taking a smoke of Peebles (Ohio) burley tobacco. One drover later remembered: "I have stayed over night with William Sheets on Nigger Mountain, when there would be thirty six-house teams on the wagon yard, one hundred Kentucky mules, in an adjacent lot, one

[32] *Franklin Farmer* (Frankfort, Kentucky), December 7, 1839; King, "Ohio Droving," 251-52; *Daily News* (Cincinnati), August 24, 1839; Henry Bushnell, *History of Granville* (Columbus, 1890), 145.
[33] Windell, "Moore's Trip," 90; King, "Ohio Droving," 249-50.

thousand hogs in other enclosures, and as many fat cattle from Illinois in adjoining fields. The music made by this large number of hogs, in eating corn on a frosty night, I will never forget." Nor were the hogs furnishing the only music: "the wagoners would gather in the bar room and listen to music on the violin, furnished by one of their fellows, have a 'Virginia hoe-down,' sing songs, tell anecdotes, and hear the experience of drivers and drovers from all points on the road."[34]

The scene was not always so rustic. At John Miller's Inn, seven and a half miles west of Hagerstown, on the heavily traveled National Road, young men and maidens of the vicinity, stagecoach drivers, wagoners, and cattlemen tripped gaily to the notes of everything from the "hoe-down" to the cotillion, winding up usually with the Virginia Reel.[35]

Dawn meant another day on the road for the outfit. About sixty to eighty days after leaving the Scioto, the herd would reach the seaboard market city. They would arrive at about that time, that is, provided unexpected trouble had not befallen them. Highwaymen infested the Wheeling, Steubenville, and Pennsylvania roads as late as 1820. If the drovers started from Illinois, they had to beware of horse thieves and cattle rustlers, who were said to have a chain of stations situated in groves across the prairie. Even if brigands did not assault the drovers, diseases in the country being passed through might strike down the cattle; sore mouth lurked in southeastern Pennsylvania in 1820, and a leg disease infested New York State cattle in 1821.[36]

Rampages of nature—snowstorm and flood—further imperiled the drive. A severe snowstorm on April 29 and 30, 1823, deposited two feet of snow on the roads leading to Baltimore. A heavy snow one foot deep in New York state on March 13, 1841, impeded droving. Swollen streams in central Kentucky

---

[34] Jesse J. Peirsol, quoted in Thomas B. Searight, *The Old Pike* (Uniontown, Pa., 1894), 142-43. On a cattle drive the drover was the employer and the drivers were the men he employed.   [35] Searight, 198.

[36] *Western Star*, May 9, September 30, 1820; June 4, 1821; Gould, "Wanderings," 90.

in March, 1830, likewise hampered the drives. John Webb, drover for Tom Shelby, wrote his boss a tale of woe from the drove stand at Bedford: "I have had the toughest time that ever a lot of cattle had . . . the debth of the mud was unbelievable . . . the 2 Day in the mountains I Had a Heavy rain . . . I have not bin able to marke more than from 5 to 8 miles per Day and Hard Driving to do that . . . thee roads east of the Alegany is dry and so Hard that it is Hard for cattle to that Has travel through so much soft roads to go on"[37] The roads to the west of Philadelphia were so muddy that there were only 416 beeves in the Philadelphia market on April 8, 1843. The opposite condition—drought—withered southeastern Pennsylvania so severely in the summer of 1841 that the local feeders refused to buy cattle; Harris kept his cattle there until the drought broke. Added to all these difficulties for the drover was the caprice of human beings; steamboat captains sometimes deliberately blew their whistles to frighten droves along the bank into taking off "as if the deuce was in them."[38]

After marketing the cattle, the drover had the problem of guarding the money belt on the way back. He and his assistant would take turns staying awake at night as guard. Harris believed that once he almost got stabbed in his sleep in a Dayton tavern, and another time almost got run through by his traveling companion in Illinois. Once some men came to Harris at Kirby's Tavern at Mt. Vernon, told him there was a settler nearby who would sell fifteen young oxen at any price in order to make good his land claim; while riding out through the "jack timber" to take advantage of this bargain, Harris got the idea he was going to be waylaid, and so he turned around. Another time, he says, he saw an armed man behind a tree.[39]

The end of driving must be attributed to the coming of the railroads, but the interesting thing is that it did not occur

[37] *Western Star and Lebanon Gazette*, May 3, 1823; *Scioto Gazette*, April 22, 1841; *Western Citizen* (Paris, Kentucky), March 27, 1830; John Webb to Thomas H. Shelby, March 12, 1847, in University of Kentucky Library.
[38] *Scioto Gazette*, April 13, 1843; Harris, "Autobiography," 77-78; Thompson, *A History of Livestock Raising*, 94.
[39] Graham interview; Harris, "Autobiography," 81.

immediately and that driving within the Middle West continued into the twentieth century. Early railroad rates were considered high by the cattlemen: "For lame bullocks that are sometimes sent from Harrisburg to Philadelphia, they charge half as much as it costs to drive them all the way—750 or 800 miles—from Kentucky to New York—the one being $8.00, the other estimated at about $16." In 1855 the fare from Columbus to New York via the Erie Railroad varied from $9.50 per head ($152.00 per car) to $12.50 ($200.00 per car). From Lexington to New York via the same route the fare varied from about $14.00 ($220.00 per car) to $18.00 ($288.00). Such changes in rates caused consternation among cattlemen, who considered $9.50 reasonable but $18.00 unfeasible. As the latter fare was higher than the cost of droving ($16.00 from Kentucky, perhaps as little as $10.00 from Ohio), drovers said it was better to sell Kentucky cattle at Louisville and Maysville for seven cents a pound. Many cattlemen at first believed that being crowded in a moving cattle car so frightened the animals that they would lose more poundage than in walking across the Alleghenies; it was thought in 1846 that after twelve hours in the cattle car, cattle must lie down, "which they cannot do in the cars like a hog."[40]

The cattlemen and drovers rarely had difficulty with railroad personnel, but occasionally they complained about the service: "[Mr. Garton of Illinois, who accompanied his drove] speaks in the highest terms of all the railroad agents . . . except him of the New York Central Railroad at Suspension Bridge. Of this, Mr. Garton, and many other drovers, think the company would make an improvement by substituting a bundle of straw upon a forked stick with a hat on. Drovers are anxious to know why cattle cars are furnished at Buffalo in abundance, while they are compelled to ship in box-cars at the Bridge."[41]

Probably because of some of the early objections to railroad

[40] *Western Farmer and Gardener*, November 15, 1846; *Ohio Farmer* (Cleveland), November 24, 1855; Jones, "The Beef Cattle Industry," 2:314. The early cattle cars had water troughs, but apparently this idea proved unsatisfactory. Goff interview. [41] *Ohio Farmer*, May 3, 1856.

shipping, Edward Sumner's cattle in the 1850's were still being driven to the eastern markets. A drove of 104 head, fattened by John Crouse of Ross County, Ohio, walked to Philadelphia in the spring of 1854 and was being driven on to New York when a speculator bought the entire drove on the road in the first week of May. That was just about the end of transmontane droving from Ohio. Even after the Civil War, however, the Goodwine family drove cattle from Warren County, Indiana, and Vermillion County, Illinois, to Chicago.[42]

Lower freight rates, better accommodation for livestock, and more branch lines all helped to destroy old prejudices and bring an end to long-distance driving east of the Mississippi. Especially notable was the opening in 1856 of a railroad between Washington Court House, in central Ohio, and Wheeling, a western terminus of the Baltimore and Ohio; via this route, cattle were shipped from Cincinnati to New York for $168 per car.[43] The driving itself had become meanwhile increasingly awkward. As towns and cities grew in population, it became more and more difficult to drive livestock through their streets. To meet this problem, the livestock markets were moved toward the outskirts of town, the New York market moving from its Wall Street site of colonial times to a location in the 1830's near 42nd Street and Broadway. But it was a losing battle and when the houses reached the Eighties north, driving to Manhattan ceased.[44]

Short-distance driving in southern Ohio continued until the advent of the motor truck. The system was essentially the same in 1900 as had been used nearly a hundred years before. A professional drover would come into, say Ross County, and associate himself with a local farmer who would take him around to farms where there were likely to be cattle available. A herd would be assembled and driven up the Scioto Valley to the railroad at Columbus.[45]

[42] *New York Daily Tribune,* May 9, 1854; Jones, "The Beef Cattle Industry," 314; Gates, "Cattle Kings," 20.
[43] *Ohio Cultivator,* March 15, 1856; Jones, "The Beef Cattle Industry," 314.
[44] Wentworth, *Sheep Trails,* 67.   [45] Cunningham interview.

How well did the driving in the first half of the nineteenth century pay? A professional drover most years could buy cattle for two to five cents per pound. The cost of driving a herd of one hundred cattle from Kentucky or Ohio to Philadelphia or New York ran from $1,200 to $1,500, depending on the actual distance traveled.[46] At the end of the trail the drover might be bankrupt or might win a small fortune, for cattle prices fluctuated sharply.

There were many factors involved in making a profit on a herd of cattle. Lying on his back in a New York hotel room on July 31, 1836, Abner Cunningham studied the patterns of light on the ceiling and thought about this business of driving cattle from Kentucky to New York. Then he wrote Brutus Clay what he thought they had done wrong: "We had some stock that was to young for to travel so far & the greatest of all we paid too much for them to have made much for our troble we started at least one month two late for this season from our country."[47]

Naturally, with prices going up and down, rumors flew along the cattle trails. A drover coming back would be questioned by a drover on his way east; if the eastbound cattleman heard that the Baltimore market was glutted, he might change his mind and take his herd to Philadelphia instead. By the time he got to market, the situation might be completely reversed. To wait outside the market city for the prices to go up was not always practicable, for the local farmers or drove-stand keepers might seize the opportunity to charge exorbitant prices for grain and pasture. Speculators often met the drover three or four days' journey from the market city to try to take advantage of him by frightening him with reports of a market glut and then by offering to relieve him of his herd at only a "minimum" loss.[48]

While the drover was worrying about vicissitudes of markets, the consumer was decrying him as a middleman who allegedly kept the price of beefsteak high. A New York agri-

[46] King, "Ohio Droving," 249-50; W. Renick, *Memoirs,* 26-30; *Western Farmer and Gardener,* November 15, 1846.
[47] Cunningham to Clay. [48] King, "Ohio Droving," 252.

cultural paper warned that if the consumer tried to boycott beef, "we will eat our beef, sell you our pork and mutton; keep our cows, and in the increased consumption of dairy products, clear from each cow a large part of what she would have brought in the city market as beef. . . . When hungry stomachs and lank limbs, are pitted against clover fields, fat oxen, and burdened granaries, there is no great difficulty in determining which must knock under."[49] But the Ohio or Kentucky cattleman was not so sure, for he was not located near 'enough to urban populations to be able to switch readily from beef production to dairying.

The effect of the transmontane drives upon the beef-cattle industry of the seaboard states was not as great as might be supposed. Connecticut farmers kept only a dwindling share of the New York City cattle market; and the New York state beef-cattle industry also declined. In fact, there was a 5.5 percent decrease in the total number of cattle in the seaboard states from the 1840 to the 1850 census. But with the beef boom of the 1850's, the number of beef cattle in seaboard states south of New York increased 17 percent. In the late 1850's New England filled up a substantial portion of its own cattle markets (about 84 percent of the Cambridge, probably not so much of the Brighton, markets), despite the fact that the railroads had by now tapped the western cattle regions.[50]

The drives seem to have had their greatest long-term effect in the Middle West, where they did much to mold the agricultural economy of the region. Born of necessity, the drives came first out of two corn-producing lowlands, the Bluegrass and the middle Scioto Valley. The drives paid well enough to be continued for half a century and to entice the farmers of other areas into the business of raising corn and fattening cattle. The success of this business stimulated the selective breeding of beef cattle in the Middle West and helped seat it in Ross, Pickaway, and Fayette counties (Ohio), where vestiges of it

[49] *Monthly Genesee Farmer*, July, 1839.
[50] Leavitt, "Livestock Industry," 42, 51, 60-61, 74, 75.

remain today.[51] Without driving, which brought thin cattle from the range to be fattened for distant markets, the distinctive feeding areas—the middle Scioto, middle Wabash, and middle Sangamon—might not have been developed. The drives from range to feeder area set the range-feeder relationship which has now come to exist between the Great Plains and the Corn Belt. Chicago in 1845 was the chief place where stockers off the range were marketed to corn farmers; today, since the Corn Belt has pushed the range westward, Kansas City is the main location of this function. Philadelphia, the destination of so many transmontane drives, has been replaced by Chicago as the greatest market for fat cattle. The expansion of the Corn Belt into areas which had been range began long before the Civil War; as feeding proved successful, it expanded from its original location on the middle courses of the Scioto, Wabash, and Sangamon into adjacent or nearby areas which had a natural geography favorable for feeding. Cattle feeding did not expand into those ranges which could not produce abundant corn, such as parts of the "Barrens" of Kentucky. In the oldest of the western cattle-fattening regions, the Kentucky Bluegrass, men turned increasingly to the breeding of fast horses and the raising of tobacco, though this has always remained Kentucky's greatest fat-cattle region as well.

[51] Of the imported cattle sold by the Ohio Company on October 29, 1836, 18 remained in Ross County, 11 went to Pickaway, 4 to Highland, and 2 to Fayette. Plumb, "Felix Renick," 27-44. The southeastern margin of the west-central Ohio upland is still a cattle and breeding center; Ohio state highway 384 all the way between U. S. Highway 22 and Greenfield is lined with breeding farms.

Chapter 6
Stockyards and Slaughterhouses

ON ITS WAY TO THE CONSUMER, OHIO VALLEY BEEF PASSED through the stockyards and slaughterhouse, finally emerging as fresh beef for the retail shops or as cured beef for export to other American cities or foreign cities. This chapter will treat the rise and location of cattle markets, the cattlemen's explorations of these markets, how the stockyards and slaughterhouses operated, the nationwide trend of cattle prices, and, briefly, the individual histories of the more important cattle markets.

Terminal markets for live cattle had not always existed in America. They do date far back into colonial times, however, having become established at Boston, New Amsterdam, Philadelphia, Baltimore, and Charleston. New England's first meat-packer was William Pynchon and Son at Boston, in 1662. The Dutch had cattle pens and slaughterhouses at New Amsterdam before that time. Philadelphia began to pack meats before 1729. The herds coming to these terminal markets were sometimes driven long distances, particularly those coming from the Carolina back country to the northern markets. Hides and tallow figured more importantly in this trade than later. Some of the beef packed in the seaboard cities was exported, as the British mercantilist system envisaged the continental colonies supplying beef, grain, and lumber to the British West Indies.[1]

On the other hand, more of the colonists lived in the country and the villages than in the cities, and the predominant economic unit was still the group of country people near the village store. The farmers killed their own cattle for local use. The village merchant handled this beef, and also pork, poultry, and other farm produce.

With the advent of the Industrial Revolution, the terminal market finally gained predominance over the country store. The real break between the two systems came with the settlement of the Ohio Valley. The change is revealed in the career of Ephraim Cutler, a merchant and land speculator at Amesville in Athens County, Ohio, who began in 1800 to export eastward the cattle that he received in payment for merchandise.[2] The transition between domestic and commercial marketing was made in the lifetime of George Renick, who in 1798 was running a general store at Moorefield, on the South Branch of the Potomac, and who seven years later as cattle king of Ohio was making the first drive of fat cattle over the mountains from that state. This transition does not imply that the people making the settlement of the Ohio Valley were less able to get along without the outside world, but rather that for the first time they could produce staple crops in real abundance and, concurrently, there was a genuine demand for foodstuff in what was coming to be recognized as a world market.[3] The Industrial Revolution had begun to concentrate the population in great centers, and wheat and meat were in great enough demand to justify long shipment. The population of New York City, for example, rose from 79,206 in 1800 to 4,174,779 in 1860.

The problem of glut, which had become occasionally and acutely in evidence by the latter date, arose when the abundance of land, the consequent abundance of produce, and the development of roads and water transportation depressed the urban markets even despite the fast-growing population. At such times the charges of long-distance transportation were more than the value of the produce could stand, and farmers lost money. They diagnosed the trouble as a lack of transportation facilities; hence many of them advocated Clay's "American System" and, later, railroads.[4]

Some Ohio Valley cattle, nevertheless, were butchered within forty miles of where they were fattened. The small-town butchers within the Kentucky Bluegrass took some of the local

[1] Charles Wayland Towne and Edward Norris Wentworth, *Cattle and Men* (Norman, Oklahoma, 1955), 142, 307-309.
[2] Julia P. Cutler, *Life and Times of Ephraim Cutler, Prepared from his Journals and Correspondence* (Cincinnati, 1890), 69,87.
[3] Benjamin Horace Hibbard, *Marketing Agricultural Products* (New York, 1923), 6.
[4] Hibbard, 7-9.

cattle to supply the local tables. Strauder Goff received this order from a small-town butcher who apparently already had the modern idea of "baby beef": "You will please pick out 24 head of them fatest cattle and let Mr. Ashbrook have them for Bonesboro pick them so as to get six of the smalest fatist cattle."[5] The rest of the local butchering was intended for shipment. Huffnagle and McCollister began butchering at Chillicothe in 1819, and barreled beef was being flatboated down the Scioto from that town and Circleville in the same year. Towns such as Chillicothe, Lebanon, Lafayette, and Quincy— all accessible to water transportation—eventually had one or two merchants who would buy dressed meats in the fall and store them in warehouses until the river or canal opened in the spring. In Chillicothe one such merchant in 1840 was Joline, who also bought stock on the hoof. By the late 1840's, Fraser's slaughterhouse in Chillicothe was packing beef for the British market. Campbell and Brown and Company began beef packing in Chillicothe in 1848, with the advertisement that they had Irish workmen well acquainted with preparing the meat for the English market.[6]

In supplying the local market and some of the export trade, the local slaughterhouses took only a minor portion of the local cattle. The major portion went to external slaughterhouses, which might be located at the periphery of important cattle-fattening regions, as in Louisville and Cincinnati, or much farther away, as in Boston and New Orleans. Some 15,000 head of cattle were taken east from the Scioto Valley by drovers in 1849.[7] Ohio cattle were commonly marketed at Pittsburgh, New York, Philadelphia, and Baltimore.

The cattlemen of the Kentucky Bluegrass looked more widely than the Ohioans for external markets—in fact, to almost every

[5] B. Finch to Strauder Goff, April 28, 1845, in possession of Ben Douglas Goff, Sr., Winchester, Kentucky.
[6] *Western Star* (Lebanon, Ohio), January 26, 1819; Theodore L. Carlson, *The Illinois Military Tract: A Study of Land Occupation, Utilization, and Tenure* (Urbana, Illinois, 1951), 89; *Scioto Gazette* (Chillicothe, Ohio), April 9, 1840; January 12, 1848; *Ohio Cultivator* (Columbus, Ohio), May 15, 1849.
[7] *Ohio Cultivator*, March 1, 1849.

point of the compass: Cincinnati and Louisville; Charleston, South Carolina, and Nashville in the South; New Orleans; Pittsburgh and the eastern markets; and even Great Britain. Although the Cincinnati market came to be regarded as a sort of private reserve by the cattlemen of the Bluegrass and of the Miami Valley, they supplied this market indifferently in the spring of 1843, bringing only fifty cattle there the week of April 20.[8] In May, 1845, B. F. Cloud, agent-drover for Strauder Goff, got Strauder an order in Cincinnati for twenty-four head of his fattest cattle; and Goff himself made frequent trips to Cincinnati, telling his friends that he could make the ninety miles on horseback in one day. The prize and second-best bullocks of the Paris fair in 1849 were sold to Cincinnati butchers.[9] Louisville was a traditional market for the Bluegrass cattlemen and one in which they had almost no outside competition. Some of the Louisville slaughterhouses sent men to the cattle regions of Kentucky to buy or contract for cattle; but Maxcy, like Neff at Cincinnati, did not buy cattle but packed beef for a fee.[10]

Beef cattle for years constituted a major part of Kentucky's overland trade with the South.[11] The normal competition for Bluegrass cattle in Nashville came from Tennessee and Christian County, Kentucky. In Charleston, South Carolina, it came from Tennessee, upland North Carolina, and the Shenandoah Valley of Virginia, although this last region tended to be oriented more toward Richmond and Baltimore. But in 1842 one Bourbon County farmer complained that Indiana, Illinois, and Missouri could supply the southern market more cheaply

[8] Robert Leslie Jones, "The Beef Cattle Industry in Ohio prior to the Civil War," *Ohio Historical Quarterly* (April, July, 1955), 64:194; *Scioto Gazette*, April 27, 1843.
[9] B. F. Cloud to Strauder Goff, May 13, 1845, Ben Douglas Goff, Sr., collection; Ben Douglas Goff, Sr., interview with the author, Winchester, Kentucky, November 8, 1955; *Western Citizen* (Paris, Kentucky), December 28, 1849.
[10] William P. Hart to Virginia Shelby, December 3, 1836, in possession of the Filson Club, Louisville, Kentucky; Thomas Maxcy to Thomas Shelby, September 11, 1845, in University of Kentucky Library.
[11] Thomas D. Clark, "Livestock Trade between Kentucky and the South, 1840-1860," *Register of the Kentucky Historical Society* (September, 1929), 27:578.

than he and his neighbors could do it. Central Kentucky cattlemen and their agents, nevertheless, very nearly trampled one another in their efforts to contract for the Charleston market in the late forties.[12]

Kentucky beef mingled with that of the whole Mississippi Valley, of course, at New Orleans, although the dangers of river transport kept some cattle from that market. Nat Hart advised his sister that shipping on flatboats was unsatisfactory and that a trip overland to South Carolina would be much safer: "You speak of sending your cattle down the River but the trouble and hazzard is so great that I feel unwilling that you should . . . , unless you could get some competant man to join you in the project. If they are shipped in flatboat it must be toed by a steamboat, and too to one it will get shoved on the way, as a flat has not the strength to stand by many dangers . . . , and when the Steam Boat gets into difficulty the flat is sacrificed to save her. The Hulk of an old Steam boat is the only safe mode."[13]

Pittsburgh, Baltimore, Philadelphia, and New York constituted a large market for fat cattle, but after 1830 the Kentuckians encountered severe competition from Ohio and eventually, perhaps, prejudice against the quality of their beef. Pittsburgh and other eastern markets were reached by overland droving, but in 1848 Tom Shelby got the idea of sending a parcel of cattle from Maysville to Pittsburgh by steamboat. Although transporting cattle on steamboats was not unusual in the Ohio and Mississippi valleys, Tom Shelby had a bad time with it. Related Shelby: "I tell you, John, if boating searves everybody like it did us with that lot of cattle, I dont think it will be done very often. I expected nothing else, but to see them look hollow; but they seemed to me to be passed hollow. I never in my life, saw a lot of cattle cut up as badly in my life." On the voyage a steer got stuck under one of the berths at the

[12] James Gay to Strauder Goff, April 28, 1848, Ben Douglas Goff, Sr., collection; Walter Chenault to Brutus J. Clay, October 23, 1846, Cassius M. Clay collection; *Western Citizen*, May 6, 1842.
[13] Nathaniel Hart to Virginia Shelby, January 2, 1837, Filson Club.

back end of the boiler room, and not the whole crew could get him out. By the time Pittsburgh was reached, he was so bruised that he was "ruined" and had to be sold for $21. Shelby had more luck, however, in getting his cattle off the boat "Every bullock took the gang-way, but 4 and they jumped overboard, but were not lost."[14]

The idea of sending the cattle alive, of course, was to supply fresh beef to distant markets. The difficulties of transporting live cattle on a boat could be avoided by packing the beef first and loading the barrels of meat on board. Of course, the boat still might sink, and if the Kentucky cattleman still owned the beef, he lost unless it was insured.

Tom Shelby, nevertheless, attempted to reach the British market. He had William Neff at Cincinnati pack some of his beef in tierces ready for shipment in the winter of 1845. This meat was consigned to a Mr. Beresford of Liverpool. A new, deep-draught ship, which was expected to make the ocean voyage to Liverpool, waited at Marietta for the ice to break up and the river to rise enough to take it over the falls, as the draught was too deep for the Portland Canal. During 1845-1847, at least, Tom Shelby regularly shipped packed beef to consignees in the British Isles via New Orleans. His shipments in the winter of 1845-1846 consisted of 598 tierces of prime mess beef.[15]

The Shakers of Pleasant Hill, in Jessamine County near the palisades of the Kentucky, also had cattle to export. Emulating the colonial New Englanders who had sent skiffs trading down the Atlantic coast, the Shakers fitted out a "trading" boat and sent it down the Kentucky River in the late fall of 1845 with the intention of trading with the people living along the Mississippi River in Arkansas, Louisiana, and Mississippi. Just below the mouth of the Red River the boat got snagged, and it

---

[14] Herbert A. Kellar (ed.), *Solon Robinson, Pioneer and Agriculturist* (Indianapolis, 1936), 2:124; Thomas H. Shelby, Jr., to John Webb, March 23, 1848, in University of Kentucky Library.
[15] William Neff to Thomas H. Shelby, December 19, 1845, February 11, 1846, in University of Kentucky Library; William Neff, Account with Thomas H. Shelby, 1845-1847, in University of Kentucky Library.

sank in fifteen minutes with the loss of all cattle except one calf. No human lives were lost, however, as the boat sank near the riverbank.[16]

Wherever the slaughterhouses were, the frosts of autumn and the thaws of spring determined just when "all beasts come to the pole-ax." Meat slaughtered in the summer would have to be rushed to the consumer's table or pickled immediately. The best season for packing was winter, although unfortunately for the butcher the cattle markets were not well filled in winter, but even then there were difficulties. A thaw or rain would set the packer to worrying that the meat would spoil. Extreme cold numbed the workers in flimsy buildings so that they could not handle their jobs. Often the supply of animals was insufficient to keep the plants going, since drovers disliked stormy weather of any kind, including snowstorms. Regardless of the season, the packer might suddenly have the opportunity to acquire several large droves at a good price, and then he would have the problem of summoning posthaste a large force of workmen.

Warm weather made slaughtering and retail marketing risky, hence it tended to depress the cattle market. When prices were low, the more daring of the packers would load up on cattle, which they then packed—and packed rapidly, because of the weather—for the export market.[17]

To kill a steer, the butcher hit it on the head with a short-handled axe. This tool, however, did its work so imperfectly that it was replaced by a hammer with a small head and a long handle. After being killed and allowed to bleed, the animal was skinned and eviscerated. Then the carcass was halved and hung up until it was in condition for packing.

The first step in packing was to rub the meat with dry salt; then it was ready to be pickled. Large vats contained pickle

---

[16] Shakers, Pleasant Hill, Kentucky, Family Journal of the East House, Filson Club.

[17] Towne and Wentworth, *Cattle and Men*, 312; *New York Daily Tribune*, May 16, 1854. Meat butchered and packed in warm weather had to be well hung, and even so, it was likely to be strong tasting, but apparently it could be marketed anyway. Edward N. Wentworth to the author, July 26, 1956.

and enough saltpeter to give the meat color and proper consistency. After twenty-four hours in the vats, when the blood was sufficiently purged out, the beef went into fresh pickle, and after a while, into a third, where it stayed until barreled for export. "Each bullock weighing say 800 lbs.," observed one packer, "will make about 3½ barrels beef half mess and half prime." Pickle and salt were put into every barrel or tierce (a barrel of 42-gallon capacity), and finally the barrel was headed. For a long voyage, two iron hoops might be added to supplement those of wood.[18]

The butchers had as much pride in their vocation as the cattlemen did in theirs. A butchers' parade, held in New York City on November 4, 1825, exemplifies this fact. Two boys marched with a banner, decorated on one side with the emblem of the profession—a knife and steel crossed; above, the pole-ax; below, on one side the saw, on the other, the chopper; in the circle, an ox and a sheep. The inscription proclaimed, "We Preserve by Destroying." On the platform of a large float drawn by six horses was a stall, at which a handsome white ox was feeding.[19]

The newspapers reported that the bystanders gave the butchers a great hand. Butchers were popular in those days of the 25-cent steak; and if beef prices went up, the consumer was inclined to blame the drover or the speculator, not the butcher.

The butchers and their friends the public, a Philadelphia newspaper reported in 1848, were about to rise up and "destroy a combination of drovers and middlemen, or forestallers, who conspire to raise the price of butcher's meat." On behalf of the cattlemen and drovers, William Renick denied the conspiracy and claimed that every superior Ohio Valley beef arriving at 750 or 775 pounds at Philadelphia had cost $57 including a dollar for loss of interest on the cattleman's money. An au-

[18] Rudolf A. Clemen, *The American Livestock and Meat Industry* (New York, 1923), 125; Thomas Maxcy to Thomas Shelby, September 11, 1845; Towne and Wentworth, *Cattle and Men*, 285-86.
[19] Towne and Wentworth, 285-86.

thority writing recently chides Renick for omitting to mention that on the transmontane drive the hogs ate the droppings of the cattle and thus represented almost clear profit except for their prime cost. There is no way of telling, however, how many of the droves had a mopping-up or cleaning-up drove of hogs in the rear. In fairness to the cattlemen and drovers, it might be said that it was absolutely impossible to determine in advance which of the markets would be most rewarding when they arrived; droving was a risky business, and even the practice of "holding back" (so unpopular with the butchers) led to disaster on more than one occasion for the drover or even the shrewd speculator. Finally, the cattle feeders could "pass the buck" too; they complained that they had to pay too much for stock cattle.[20]

Every kind of cattle came to market—pureblood Durhams, grade Durhams, commons, grass-feds, slop-feds. Even Texas or Indian Territory Longhorns appeared in the New York market in the 1850's. Only in the case of the Philadelphia butchers do we have record of anyone complaining that the butchers failed to pay a premium for superior cattle; otherwise, it was only in times of great demand or scarcity that the butchers paid almost as much for inferior as for superior cattle. When prices were very high, as in the spring of 1855, farmers rushed everything to market, including young beeves that had not had their customary finishing; this, of course, was a waste of potential beef production. Rare was the drover or salesman who could put one over on the butcher or buyer. References to "watered stock" appear nonexistent. Grass-feds could not be passed off as corn-feds. When Cherokee, Choctaw, or Texas Longhorns started appearing, the drover or salesman occasionally tried to pass them off as Iowa stock, but seemingly this misrepresentation was too transparent. Filing off the age wrinkles on the horns may have been tried. Slop-feds brought inferior prices and apparently were subject to considerable shrinkage; one

[20] William Renick, *Memoirs, Correspondence, and Reminiscences* (Circleville, Ohio, 1880), 26-30; Jones, "The Beef Cattle Industry," 192; *New York Daily Tribune*, June 21, 1855.

man bought a herd of slop-feds at the Belvidere, Illinois, distillery and then discovered that after six weeks on hired pasture they had shrunk 100 pounds per head—and his pocket had shrunk $20 per head![21]

A variety of factors affected prices in any given cattle market. If a storm struck during the market day, it favored the sellers, because the buyers hurried their purchases. Otherwise, wet weather depressed prices. Strawberry season lessened the demand for meat and reduced cattle prices. When no sheep came to market, cattle prices went up. When demand was stronger than supply in the market, sellers not only got their own prices, but their own estimates of weights (in the 1850's sellers and buyers weighed the animal on scales and then estimated what the weight would be not counting hide and tallow); thus cattle bought at 12 cents per pound might turn out to have cost $12\frac{1}{2}$ cents or 13 cents.[22]

Although these day-by-day local factors were important to the cattleman arriving at market, the national average of cattle prices ran in big trends which were responsive to the rest of the agricultural economy and national economy. Cattle prices were comparatively low between 1788 and 1791, rose in company with other prices in the early 1790's to a peak in 1797, then subsided a bit. Comparatively little variation occurred between 1800 and 1814. Provisions were slow to respond to wartime influences, but finally reached high levels after the war, peaking in 1817.[23]

A correlation existed in the period 1802-1817 between receipts at land offices and the prices of western agricultural commodities. Each reflected a spirit of speculation. Higher commodity prices made land a more attractive investment. A large sale of land probably meant a large number of settlers, who at first would be dependent on others for provisions, hence

[21] New York Tribune, July 11, 1854.
[22] *Ohio Farmer* (Cleveland), May 3, June 28, 1856; *New York Daily Tribune*, May 3, 10, 1855.
[23] Thomas Senior Berry, *Western Prices before 1861* (Cambridge, Massachusetts, 1941), *Harvard Economic Studies*, 74:217.

enlarging the local cattle trade. A heavy sale of land caused larger federal deposits in western banks, which thereupon were able to make more liberal loans for land purchases, commodity speculation, and purchase of stock cattle.[24]

Western prices from 1823 to 1835 were in close synchronism with the annual totals of domestic bills of exchange bought by the Bank of the United States. The percent of the bank's total exchange bought in the West and Southwest fluctuated sympathetically with prices. Thus it rose in 1825, declined to 1827, rose until 1832, then fell off in 1833 and 1834. Cattle prices rose from 1835 to 1838. The year 1838 was marked by high prices for cattle, but also, the feeders claimed, by high costs in their operation.[25]

In 1839 cattle and other agricultural prices responded to the depression spreading in the wake of the Panic of 1837, and the agricultural depression of the forties began. In 1841 occurred an apparent contradiction of an economic law. Western currency was depreciating to the greatest extent in respect to specie and the currency of the East and South, but the downward rush of western prices was delayed to only a minor degree. Whereas New England was turning from beef cattle to dairying, the Ohio Valley appears to have increased its production of beef in the early 1840's; sheep, suddenly popular with some farmers, made only local inroads, but would probably have done more had it not been for the lowering of the wool tariff in 1842. The increased production of beef did not have a salutary effect on prices. The national average of beef-cattle prices was lower in 1842, 1844, and 1845 than in any other year in the whole period 1840-1860. The Whig tariff of 1842 apparently helped produce a small general rally in 1843. Ohio's bank law of 1844, which increased the supply of paper money, at first inflated the prices of Ohio cattle; but when buyers consequently went elsewhere, notably Kentucky, which by now had the soundest banking of the Ohio Valley states, prices went down. The national index of cattle prices swung upward

[24] Berry, 373-74.
[25] Berry, 429; *Monthly Genesee Farmer* (Rochester, New York), July, 1839.

in 1846 and 1847, but receded in 1848 to the level of 1841. The upward trend was resumed in 1849, when prices reached the high of the decade.[26]

There was a little slipback in 1850 and 1851, but the next year the national effect of the California Gold Rush was felt. Thereafter an alltime high level was reached and maintained for the rest of the decade, with the average national price of cattle in the 1858 depression still being higher than any year from 1840 to 1852 inclusive. Besides the California demand, other factors encouraging this beef boom were the repeal of the British navigation laws, the demand accruing from railroad construction, and the marketing revolution effected by railroads.[27]

Following is the nationwide index of prices of beeves, good to prime, live weight, years 1840-1860 (1860=100).[28]

Year	Index	Year	Index	Year	Index	Year	Index
1840	65.8	1846	71.1	1851	76.3	1856	110.5
1841	60.5	1847	76.3	1852	89.5	1857	115.8
1842	57.9	1848	60.5	1853	105.3	1858	94.7
1843	63.2	1849	81.2	1854	97.4	1859	100.0
1844	57.9	1850	78.9	1855	105.3	1860	100.0
1845	57.9						

Because more Ohio Valley cattle were marketed at New York than at any other city, I shall attempt to trace the general trend of beef-cattle prices there over the years. Later, in discussion of other cattle markets, mention will be made of some times when New York, Philadelphia, and Baltimore had different prices. Although prices can sometimes be explained only in terms of crowd psychology, I shall attempt to explain the actions of the markets wherever possible.

All prices to be mentioned are the price per 100 pounds net weight. The net weight did not include the "fifth quarter"—

[26] Berry, *Western Prices*, 465; Charles T. Leavitt, "The Meat and Dairy Livestock Industry, 1819-60" (dissertation, University of Chicago, 1931), 43, 51; William H. Aldrich, "Report on Wholesale Prices and Wages," *Senate Reports*, 52 Cong., 2d Sess., 1892-1893 (Washington, 1893), 1:103; *Hunt's Merchants' Magazine and Commercial Review* (New York), April, 1844.
[27] Aldrich Report, 1:103; Jones, "The Beef Cattle Industry," 313.
[28] Aldrich Report, 1:103.

hide and tallow. Thus, since each pound of net weight consisted mostly of consumer beef, the price per 100 pounds net weight was higher than the price per 100 pounds "on the hoof." The price was higher, but the hide and tallow were free to the buyer. The estimated net weight was, at best, little more than half the weight "on the hoof," since most cattle "killed" 50 to 62 percent, that is, contained 50 to 62 percent usable beef. Only at Boston, where the cattle were weighed on scales, were prices quoted "on the hoof." Apparently the cattlemen preferred to have their cattle weighed "on the hoof" rather than estimated "in the quarter" (estimated net weight). There was no safe way for professional drovers to buy cattle in the Ohio Valley except on scales, counting the weight they would sell for at New York at just one half (to be on the safe side) the number of pounds they weighed alive; but the catch was that only a few of the farmers even by the early 1850's had scales, and hence the drovers could not conveniently weigh the cattle.[29]

Over the years New York City had several cattle markets. The earliest one that Ohio Valley drovers ever visited was the Wall Street market, actually located on the east side of Pearl Street between Pine and Wall, founded in 1676 and lasting into the 1830's. By that time another market was the Bull's Head, on the Hudson River at the foot of Robinson Street. Later known as Chamberlain's Bull's Head, this cattleyard and the adjoining Bull's Head Tavern lasted through the half century, although the tavern narrowly escaped the torch during the draft riot of 1861; eventually both the cattleyard and the tavern had to make way for the Bowery Theater. By the 1830's congestion in lower Manhattan necessitated the founding of still another cattle market, this at 4th Avenue and 34th Street; it had three names: Allerton's Bull's Head, Upper Bull's Head, and Washington Drove Yards.[30]

[29] *New York Daily Tribune,* May 24, 1855; *Ohio Farmer,* January 30, 1858. Two cattlemen who did have scales in the early 1850's were Tom Shelby and Brutus Clay. A scaleworks was in business in Cincinnati in 1852.

[30] Towne and Wentworth, *Cattle and Men,* 309.

The first cattle to appear from the Ohio Valley were, so far as we know, a drove of 200 head brought from Chillicothe in 1817 by a man named Drenning. The hundred head in the drove sent by George Renick and Joseph Harness in 1818 were valued when at home at $52 each and sold at the Bull's Head for $69 each, or presumably about $12 per 100 pounds. The same year, Daniel Drew's big herd of Ohio Valley cattle brought "up over $30 [profit] on every head." William Renick, finding the Philadelphia market glutted in 1824, decided to send his drove on to New York City; taking the Sunday afternoon mail coach, he reached New York about sunrise Monday morning, and, stopping at the Bull's Head Tavern, was met at the door by Daniel Drew, who immediately asked him what he would have to drink and highly recommended his Jamaica rum.

Through the 1820's, nevertheless, Ohio Valley cattle made only sporadic appearances in the New York City market. Most of the cattle came from Connecticut, some from upstate New York. The Wadsworth outfit of the Genesee country preferred to market their cattle in New York City, partly because Ohio competition was less there than in Philadelphia.[31]

When Ohio Valley cattle arrived in large numbers in the early 1830's, the New York market apparently suffered a glut. Whereas Ohio cattle in 1818 had presumably fetched about $12.00 per 100 pounds in New York, the prices in 1832, 1834, and 1835 (when times were at least as good as in 1818) were only $4.00 to $7.50.[32] The low cattle prices of 1834 coincided, as has been mentioned, with the small purchases of domestic bills of exchange in the West by the Bank of the United States.

[31] James W. Thompson, *A History of Livestock Raising in the United States, 1607-1860* (Washington, 1942), 94; W. Renick, *Memoirs*, 97, 98; Bouck White, *The Book of Daniel Drew* (New York, 1911), 81; Leavitt, "The Livestock Industry," 60; Neil Adams McNall, *An Agricultural History of the Genesee Valley, 1790-1860* (Philadelphia, 1952), 136.

[32] *Scioto Gazette*, April 4, December 19, 1832; *Mercury* (New York), October 23, 1834; September 24, 1835. Prices of $5.50 to $12.50 at New York on March 28, 1833, reflect the fact that the many cattle that had been feeding in the Scioto Valley during the winter had not yet started to arrive. *Scioto Gazette and Independent Whig*, April 10, 1833.

Then, in 1836, the market turned sharply upward. In the early winter months prices of $6.00 to $9.00 prevailed, and in the spring (when Ohio Valley cattle made their first appearance) prices for top quality beeves got as high at one time as $12.00 per 100 pounds. The New York cattle market was in general higher at this time than the Baltimore. On June 6, 600 of the 900 cattle were from Ohio; prices sagged a bit so that 800 went for $7.00 to $10.50 per 100 pounds. It was the beginning of the seasonal glut.

Arriving with his Kentucky cattle in July, Abner Cunningham encountered this sag. The market, he reported, had averaged only $8.00 for the last two weeks. He made "some little money" on the first three lots, but the last lot was sold at a small loss. "It is my opinion," he wrote, "that the next one will be [?] hard to pay for itself." Cunningham recognized his error in arriving at market in July instead of May. In late summer, New York state grass-fed cattle normally began to flood the New York market, and Ohio Valley cattlemen and drovers learned to avoid New York at this season. On October 10, 1836, the market had 1,300 cattle, many of them inferior; the average price was $7.00.[33]

The Panic of 1837 did not, as we have observed, immediately depress western agricultural produce; beef-cattle prices, in particular, remained high in 1838. At the New York cattle market they were still high on March 4, 1839 ($9.00 to $13.00 and a few $15.00). Cattle prices at New York were kept high that spring by the operations of a great cattle speculator, Steinburger, who had a controlling influence and whom Solon Robinson later called "the greatest beef speculator that ever operated in this country."[34] Prices presumably went down that summer, however.

Prices at New York made a slight recovery in the spring of 1841 but failed to participate in the moderate nationwide rally

[33] *Mercury* (New York), January 21, 28, April 6, 14, 21, May 12, 19, June 9, 23, October 20, 1836; Abner Cunningham to Brutus J. Clay, July 31, 1836, Cassius M. Clay collection.

[34] *Mercury*, March 7, 1839; *New York Daily Tribune*, May 30, 1854.

of cattle prices in 1843. Tom Shelby met disappointment when he arrived in New York in early May, 1843. There were 600 to 800 cattle in market, at prices of 7 to 8 cents per pound. "But for the arrival of 300 River Cattle from Albany," he lamented, "we should have met a fine market." He and Webb got an average of $54 per head, "which is about 7¢. We expect the present price will be sustained this [May 14] and next week but after that, it must come down, as the great glut of Ohio and Kentucky cattle are pushing."[35]

The years 1844 and 1845 marked the second pit of the beef-cattle depression of the forties, with prices on the national average as low as they had been in 1842. Prices in the New York cattle market were just as bad as, or worse than, the national average of beef-cattle prices. On May 3-4, 1844, Zell sold thirty Kentucky cattle for 5.4 or 6 cents. On May 11 there were 1,200 head at market: "The market is exceedingly depressed, oweing to the supply of this Country Cattle not yet being exhausted, and the Ohio Cattle comeing too much in a crowd." To make matters worse, Patterson's cattle from Kentucky were also at New York. Webb's lot brought about 6 cents. "If we had been two weeks later," opined Shelby, "it would have been greatly in our favor."[36] This, of course, was the exact opposite of the conclusion that Cunningham had reached eight years before; each man, of course, may have been correct in his own year, for the best time to arrive at the New York market in one year might not be the best time in another year, and no one could tell until he got there.

The depression in beef-cattle prices at New York, as in the nation at large, continued through 1845, with prices at New York in May being $5.00 to $7.00 per 100 pounds. These, nevertheless, were better than Baltimore prices; consequently, beef cattle from the South commenced to appear in the New York market. There were 650 head on May 19. Once having gained

[35] Aldrich Report, 2:25; Thomas H. Shelby to Isaac Shelby, May 14, 1843, University of Kentucky Library.
[36] Thomas H. Shelby to M. C. Shelby, May 4, 1844, Thomas Shelby to Isaac P. Shelby, May 11, 1844, in University of Kentucky Library.

a foothold, southern cattle were sent to New York throughout the rest of the forties, and from 1849 through 1852 constituted perhaps one-third to one-half of the cattle received there.[37]

Prices edged forward in 1846 at $5.75 to $7.75 and surged forward in 1847 at $7.00 to $9.00, part of the national recovery of beef-cattle prices. The latter figures prevailed despite the presence of 1,000 beef cattle in market (May 3), of which 700 had been attracted from the South. On May 8, a Kentucky cattleman wrote home from New York: "we calculate on a strong demand from Boston and all the villages for 30 miles around, the resourcers of this country. Cattle are all exhausted, and if we dont crowd the market with Western cattle it will be sustained at the present prices until August. The season has been unusually cold and backward, and nothing eatable can be produced here from grass until midsummer." In a move to take advantage of this situation, the South sent up so many cattle that on July 12 all 1,100 cattle at market were southern. Prices, as might be expected, reacted to the copious supply, slumping at $6.00 to $7.00.[38]

In 1848 some of the advance in beef-cattle prices since 1845 was lost; the New York market, strong in April, weakened shortly thereafter. At market on May 29 were 1,300 head, "all from the South and West, via Philadelphia"; the average of prices—$5.00 to $8.00—was wider than usual. The seasonal decline in the fall sent prices down at 5 to 7½ cents.

The year 1849 saw an end to the beef depression. The New

[37] Thomas Shelby to Isaac Shelby, Jr., May 5, 1845, in University of Kentucky Library; *Weekly Herald* (New York), May 24, 1845; June 23, 1846; *Scioto Gazette,* July 21, 1847; *New York Herald,* May 8, 1847; May 30, June 6, December 5, 1848; April 17, 24, May 15, June 26, July 10, 31, November 21, 1849; *New York Weekly Tribune,* April 20, 1850; December 13, 1851; May 8, 20, July 3, December 4, 1852. The newspapers record the arrival of no Ohio Valley cattle in the late spring and summer of 1849, yet it is known that some 10,000 such cattle passed through Morristown, New Jersey, during that period. Therefore, it must be concluded that some of the cattle listed as "southern" were actually Ohio Valley. I made an allowance for these in estimating the proportion of southern cattle in the market.

[38] *Weekly Herald,* June 23, 1846; May 8, 1847; Thomas H. Shelby to M. C. Shelby, May 9, 1847, typewritten copy in University of Kentucky Library.

York market, however, did not show as good a recovery as the national average; prices ranged from $5.00 to $7.00 for inferior and from $7.75 to $9.25 for superior cattle, the latter figure being reached on May 14. Nevertheless, these were better prices than could be obtained in the western cities.[39]

Prices of all agricultural commodities, including cattle, at New York City rose steadily for the next five years. A peak was reached on May 9, 1855, when $15 was the prevailing price for superior quality, and the average of all sales was $14 to $15! This, nearly four times the price of nine years before, was the highest price in the history of the New York cattle market, and one-half the cattle were in the hands of the regular cattle brokers as owners. Now was one of those occasional times in the 1850's when the price differential between coarse cattle and fine corn-fed Durhams narrowed to almost nothing. The wild prices were attributed to a national beef scarcity, which had been caused by the drain to California and by the drought of 1854. Moaned a New York City paper editorially: "Choice cuts of beef will be retailed this week at 25c a pound, and how long those who buy to eat and eat to live can bear this, remains to be seen." Relief for the consumer came fast, although by all normal standards the market remained good. By June 20 prices had receded at $8.00 to $11.00. Warm weather, which might cause meat to spoil, was depressing prices. Inexperienced buyers, emboldened by the high prices of May, had paid Ohio Valley farmers too much for their cattle and now were stuck, finding that "holding back" did no good.[40]

Early in 1856 it looked as though the crazy market might return. On April 16 prices of $10.00 to $16.00 were quoted, and fevered buyers noted an appearance of scarcity as several yards emptied before noon. There had been an unexpected demand for the Boston market, and 700 head that had been on their way to New York had been sold at Albany.

[39] Aldrich Report, 2:24, 25; *New York Herald*, May 30, June 6, December 5, 1848; April 17, 24, May 15, June 26, July 10, 31, November 21, 1849.
[40] *New York Daily Tribune*, May 10, 12, June 21, 1855.

By May 27, however, prices had receded at $9.00 to $10.50, and by June 11, at $8.00 to $10.00, and disappointed speculators began conspiring. Some of them went to meet the trains at Bergen Hill or Albany. By June 18 the speculators had ownership of nineteen droves. In all, twenty-nine droves in the market were being "held back" in a partially successful attempt to effect a price advance: prices went up about $1.00, at $9.00 to $11.00.

Nevertheless, prices thereafter slipped back, off about $1.00 to $2.00 from the same time the year before. This happened despite the fact that the national average of cattle prices was higher in 1856 than in 1855. Again in 1857 the New York market performed less well than the national average, which reached the highest point since the flush days of 1837. The New York prices of $10.00 to $12.50 were considered low, although, of course, by the standards of the 1840's they were very high.[41]

The next year, 1858, the New York market reflected the depression following the Panic of 1857. Prices fell at $7.00 to $9.00. But recovery was prompt, and one authority makes too much of a sale of Ohio cattle in New York in May of 1859 at $4.00. Actually, on May 11, the quotations were $9.50 to $12.50, and the New York market for the year 1858 was not as low as it had been for any year from 1840 through 1851. What is significant, however, is that the New York market, although almost always higher than Cincinnati between 1856 and 1859, was not as high as some other markets. Apparently the railroads and the flood of increasingly good Illinois cattle were equalizing the cattle markets. Then, in 1861, cattle prices in the West plummeted, and the New York market regained its attractiveness. The advent of the war that summer created, of course, a nationwide artificial inflation of cattle prices.[42]

[41] *Ohio Farmer*, May 3, June 7, 28, July 12, 1856; May 23, 1857.
[42] *Ohio Farmer*, May 29, November 27, December 18, 1858; May 28, 1859; Jones, "The Beef Cattle Industry," 317-18; Aldrich Report, 1:103, 2:25; Henry Ellis White, *Wholesale Prices at Cincinnati and New York* (Ithaca, New York, 1935), Cornell University Agricultural Experiment Station, *Memoir*, 182:9-12.

The New York market in typical years had heavy receipts in July and August. This is only partly explained by the arrival of the first New York state grass-feds in August. The rest of the explanation is that many Ohio Valley cattlemen were tardy in getting their droves on the road in the spring and did not reach New York until July. A high peak of receipts at New York was reached usually in September or October, and receipts for each of the remaining months were slightly higher than for either July or August.[43]

After 1852, Ohio squeezed the South out of the New York cattle market. That year Ohio Valley cattle were in market even in fall, the season of the New York state grass-feds; and by 1856 Ohio had smothered the New York state beef-cattle industry. The Ohio Valley retained its preemption of the New York City market for the rest of the decade, but now Illinois, not Ohio, formed the greatest Ohio Valley representation. In 1862 half the 165,000,000 pounds of beef dressed in New York City had come from Illinois, including parts of Illinois north of the Ohio Valley.[44]

Only a little Ohio Valley beef reached New York City already packed via the Erie and other New York canals and the Hudson River. In 1837, 54 barrels of beef came down the river to Gotham; in 1840, 7,000; in 1843, 47,000; and in 1845, 67,000. Then this commerce leveled off. The export figures at Cleveland indicate that the Ohio Valley contributed but little of this beef.[45]

Besides the usual shuttling of cattle from one market city to

[43] *Niles' National Register* (Baltimore), February 7, 1846; *Hunt's,* June, 1854.

[44] *New York Daily Tribune,* May 9, June 20, July 3, 1854; May 3, 10, June 21, 1855; *Ohio Farmer,* June 30, July 14, November 24, 1855; May 3, June 7, 28, July 12, November 29, December 5, 1856. Origin of the cattle and average weekly receipts at New York in 1856 were: Ohio, 854; Illinois, 529; New York, 503; Kentucky, 295; Indiana, 170; Virginia, 59; Iowa, 26; Texas, 22; and New Jersey, 5. *Scioto Gazette,* January 8, 1857. By 1860, however, Illinois outranked Ohio; and in the three-year total, 1857-1860, Illinois also outranked Ohio. *Wisconsin Farmer* (Milwaukee), February, 1861; Clarence P. McClelland, "Jacob Strawn and John T. Alexander, Central Illinois Stockmen," *Journal of the Illinois State Historical Society* (June, 1941), 34:208.

[45] *Hunt's,* June, 1841; March, 1844; March, 1847; January, 1849.

another, New York City exported some cattle alive to Bermuda, which was, in fact, the only noteworthy market for New York City's export of live cattle. On May 29, 1848, for example, eighty head, possibly of Ohio Valley origin, were sold for embarkation to that destination, and on May 3, 1852, forty head for the Royal Naval station.[46]

Most of the overseas market and coastwise trade, of course, had to be supplied with barreled beef, and of this, New York and New Orleans were the greatest shippers. In 1841 New York exported 21,000 barrels of pickled beef; in 1844, 61,000; in 1851, 40,147; in 1854, 95,513; in 1858, 76,643. Export was greatest in November, presumably because in that month the flood of cheap New York state grass-feds coincided with the advent of the first freezing weather which was favorable for slaughtering.[47]

Philadelphia, biggest city on the seaboard in colonial times, early became an exporter of the farm produce of the breadbasket of America, Pennsylvania. One item of export was beef; thus, in 1796, Philadelphia exported 6,860 barrels.[48] Despite, however, the nearby supply from Pennsylvania and perhaps Maryland and Virginia, early Ohio Valley cattlemen found a market here. Whether any Kentucky cattle had arrived here before Felix Renick's drive of Ohio cattle in 1815 is unknown, but it seems probable, since Kentucky cattle are known to have been going to Virginia and Baltimore, and some might logically have been taken from these places to the Quaker City.[49] In 1816, Thomas Goff got $77 a head for his 130 Patton beeves which he drove to Philadelphia from Clark County, Kentucky. Of the herd of 100 cattle which Felix Renick drove to Philadel-

[46] *New York Herald*, June 6, 1848; *New York Weekly Tribune*, May 8, 1852; July 3, 1854.
[47] *Hunt's*, April, 1844; February, 1845; January, 1849; March, 1855; February, 1859.
[48] *Hunt's*, January, 1855.
[49] Felix Renick Account Book, in possession of Renick Cunningham, Chillicothe, Ohio; F. A. Michaux, *Travels to the West of the Allegheny Mountains*, in Reuben Gold Thwaites (ed.), *Early Western Travels* (Cleveland, 1904-1907), 3:245; *American Farmer* (Baltimore), June 29, 1821.

phia in 1817, 20 of the best he sold for $160 per head, the whole lot averaging $133.[50]

Once Ohio cattle began coming regularly in the 1830's, the Philadelphia market became an object of continuing attention by Ohio cattlemen.[51] It is not true, as one authority says, that the feeders of southeastern Pennsylvania presently blocked Ohio out of the Philadelphia market, although in contrast to its near-dominance of the New York market, Ohio maintained little more than a good foothold at Philadelphia. Like Ohio, the South did not make as deep a penetration into the Philadelphia market as at New York. At times, New York state grass-feds, overflowing to Philadelphia, made a deep penetration. Kentucky put in only occasional appearances at Philadelphia, as when Finch sold three lots of Kentucky cattle here in July, 1836.[52] Some of the cattle sent up from Baltimore or from Virginia feeders, however, were probably of Kentucky origin.

The Philadelphia cattle market was the Union Drove Yards; the drovers' hostelry, the Barley Sheaf Hotel. The city developed a favorable balance of trade with the West. It accumulated drafts on New Orleans and New York, and like New York, it handled many of New Orleans' business transactions abroad.[53]

Philadelphia prices showed the same trend as New York but were not always at the same level. William Renick charged that the Philadelphia butchers stubbornly refused to pay a worthwhile premium for superior cattle, hence many of these passed on to New York City. In October, 1836, prices at Philadelphia were no more than 50 cents per 100 pounds better than the prevailing prices for grass-feds at New York City. In

---

[50] Lucien Beckner, "Kentucky's Glamorous Shorthorn Age," *Filson Club History Quarterly* (January, 1952), 26:39; Charles S. Plumb, "Felix Renick, Pioneer," *Ohio Archaelogical and Historical Publications* (January, 1924), 32:21.

[51] See, for example, *Scioto Gazette*, April 20, 1843.

[52] Jones, "The Beef Cattle Industry," 194; *Scioto Gazette*, July 1, 1841; May 11, 25, June 15, November 2, 1843; May 2, 23, June 13, July 21, 1844; *New York Herald*, June 4, 1847; April 15, December 2, 1848; Abner Cunningham to Brutus J. Clay, July 31, 1836, in possession of Cassius M. Clay, Paris, Kentucky.

[53] *Hunt's*, April, 1844.

early May, 1843, prices at Philadelphia were $4.50 to $6.50, but at New York, $7.00 to $8.00; accordingly, on May 4, 183 of the 369 Ohio cattle at Philadelphia were taken off the market by their owners or salesmen and moved to New York. In early May of the next year the New York and Philadelphia markets were in exact synchronism, but later in the month the price of superior cattle at New York pulled ahead of Philadelphia by about $1.00. June of 1847 saw prices of $6.50 to $8.00 at Philadelphia, but $7.00 to $9.00 at New York; hence, 300 of the western cattle at Philadelphia on June 3 were taken to New York, arriving, the handlers hoped, before the upstate grass-feds. Some 650 grass-feds were taken from New York City to Philadelphia apparently to take advantage of a fifty-cent differential in the price of inferior cattle. On May 10, 1849, Philadelphia had 1,400 cattle, of which 900 were sold at $6.25 to $7.25 and 400 were taken to New York; four days later New York had 1,500 cattle and prices of $7.00 to $9.25.[54]

The first Ohio fat cattle to be driven over the mountains were marketed at Baltimore in 1805 by their owner, George Renick, who made an average profit of $31.77 per head.[55] Thereafter, Baltimore was the third most important eastern market for Ohio Valley cattlemen.

In Baltimore the market reached the saturation point more quickly than in New York; and it was no more closely synchronized with Philadelphia and New York than were the latter two cities with each other. Thus, on April 18, 1836, the price of good cattle at New York was $10.50, but at Baltimore two days later, $9.00; and in the first week of June at New York, $10.50, but at Baltimore, $8.50. Baltimore "bottomed out" in 1840, a depression year; 1,000 cattle glutted the market on October 13 and only 400 were sold, and those at the ruinous

[54] W. Renick, *Memoirs*, 26-30; *Mercury*, October 13, 20, 1836; *Scioto Gazette*, May 11, 1843; May 2, 23, 1844; July 21, 1847; May 16, 1849; *New York Herald*, May 8, June 4, 1847; May 15, 1849; Thomas H. Shelby to Isaac Shelby, May 14, 1843, May 11, 1844; Thomas Shelby to M. C. Shelby, May 4, 1844, May 9, 1847, in University of Kentucky Library.

[55] J. F. King, "The Coming and Going of Ohio Droving," *Ohio Archaeological and Historical Publications* (April, 1908), 17:249-50.

price of $2.25 to $3.00. On June 25, 1841, a drove of 120 head was taken north; but the owners met disappointment, for the next week in Philadelphia, prices were lower ($5.50 to $7.50) than in Baltimore ($6.50 to $7.25). Small supply did not necessarily insure good prices at Baltimore unless the beeves were good; on August 10, 1847, 300 head at market went for $4.80 to $6.75, while on the same day at New York, where a large drove had come down the North River and sent receipts up to 1,600, some sales were as high as $7.00. On April 13, 1849, Baltimore prices were $3.50 to $7.75; 408 beeves were driven to Philadelphia, where prices on the same day were $6.50 to $8.50, or close to the New York prices of $6.00 to $9.00. On May 7, with Baltimore prices at $5.00 to $8.50, 200 head were driven to Philadelphia, where, however, prices were giving way at $6.25 to $7.25; hence 400 beeves were taken from Philadelphia to New York, but whether they arrived there in time to benefit from the May 14 prices of $7.00 to $9.25 is not known.[56] One can only conclude that the profitable marketing of cattle demanded a good deal of luck.

The fact that Baltimore could absorb fewer cattle than either Philadelphia or New York did not necessarily make it a poor place to sell. Baltimore butchers, nevertheless, by necessity or choice, went out themselves to get beeves. The agent of a Baltimore butcher paid Strauder Goff $3.00 per 100 pounds for fat cattle delivered at Cincinnati in late 1843, Goff thus saving the expenses of the transmontane drive, and sparing himself worries about the vicissitudes of markets. Other Baltimore butchers' agents went to Clark County, Kentucky, and assembled herds there which they held in waiting until they knew the sailing dates of ships from Baltimore.

Baltimore had exported only 572 barrels of beef and forty-two live cattle in 1842, but by 1847 the annual export was nearly 10,000 barrels. Of this, nearly half went to Great Britain.

[56] *Mercury*, April 21, 28, June 9, 1836; *Niles' Register*, October 17, 1840; *Scioto Gazette*, July 1, 9, 1841; August 18, 1847; April 18, May 16, 1849; *New York Weekly Herald*, May 24, June 7, 1845; *New York Herald*, November 30, December 2, 5, 1848; April 17, 24, May 15, 1849.

In 1854 the export was just under 6,000 barrels and tierces; two years later it was about 7,500.[57]

With the possible exception of St. Louis, New Orleans was the only market reached not by overland driving, but by water. In the beginning, vessels of all kinds tied up wherever they could find a convenient place before the quay of the French city, and later the flatboats gathered along the Tchoupitoulas Street sector in the American Quarter above Canal Street. When steamboats appeared in the decade after 1810, they were ordered to dock only between the upper side of Canal Street and the lower side of Customhouse Street, a distance of about three blocks.[58]

Ohio Valley beef came downriver to New Orleans either alive or barreled, and either way it might come on flatboats, steamboats, or hulks. The cattle or the barrels of beef being disembarked at Orleans might or might not still be in the ownership of the Ohio Valley cattleman. In the early years they usually were, and in fact the man who had fed the cattle often came with them, or with the barrels of beef, on the flat, which also probably carried pork and whisky. Practically all business was handled on a personal basis; there was no central market for anything, not even cattle, and sellers had to hunt buyers. During the steamboat era it was less common for the cattleman to come with his cattle or barrels of beef (which might have been packed for a fee by someone like Maxcy at Louisville); William P. Hart, for example, was able in the 1840's to hire a hand to take a parcel of Virginia Shelby's cattle to New Orleans for $25 per month. Another way the Ohio Valley cattleman could market his stuff at New Orleans was to consign his shipment to an agent in New Orleans, possibly the steamboat captain, who for a commission would find a suitable buyer. This, of course, was the same function performed by the "cattle brokers" or "salesmen" at New York's Washington Drove Yards in the 1850's, but at New Orleans the forwarding

[57] Strauder Goff to J. C. Hughes, August 28, 1843, Ben Douglas Goff, Sr., collection; Goff interview; *Hunt's*, June, 1843; August, 1848; April, 1858.
[58] Harold Sinclair, *The Port of New Orleans* (New York, 1942), 166.

merchants and commission men handled many other items beside meat. In a few instances, as we have mentioned, the cattle or barrels of beef arriving at New Orleans were no longer in the ownership of the Ohio Valley cattleman. A Louisville commission man like Bronson Wheeler may have already negotiated the sale to a New Orleans butcher or shipper; or a speculator may have bought the shipment outright at Louisville from the cattleman.[59]

Probably there had not been enough beef cattle in Kentucky (then part of Virginia) in the 1780's to make beef an article of export to New Orleans at that time. It seems reasonable to suppose, however, that by about 1795, which was approximately when cattle driving to the East started, Kentucky had some beef to export, and this must have gone via flatboats to New Orleans. At about the beginning of 1811, flatboats and keelboats departed downstream from Louisville with 6,300 pounds of beef during a two-month period. Beef was reported "very precarious and of little credit" at New Orleans on December 14, 1817. By June, 1819, prime beef was bringing $12 a barrel on the New Orleans wharves and was dull at that. In mid-February of 1820, taking advantage of the first rise of water in the Scioto since the previous spring, thirty to forty flatboats left Chillicothe for New Orleans. No doubt these Ohioans met disappointment at New Orleans, for on February 25 word was sent from New Orleans that Ohio's country produce could "not be sold at any price." In fact, sales could not always cover even freight charges. The same issue of a Miami Valley newspaper that brought this sad news also recorded the hazards of the New Orleans voyage: four or five flatboats were reported sunk by ice near the mouth of the Wabash, and the *St. Louis* was wrecked eight miles above New Madrid.[60]

[59] Sinclair, 169; William P. Hart to Virginia Shelby [ca. 1842-1847], Filson Club; Louis C. Hunter, *Steamboats on the Western Rivers* (Cambridge, Massachusetts, 1949), 348-49; Bronson Wheeler to Brutus J. Clay, February 5, 1847, Cassius M. Clay collection.

[60] Zadok Cramer, *The Navigator* (Cincinnati, 1814), 360; *Lebanon Farmer* (Lebanon, Ohio), January 24, 1817; *Springfield Farmer* (Springfield, Ohio), March 17, 1819; *Western Star*, February 22, March 28, 1820.

The shippers and commission men at New Orleans had little sympathy for complaints about low prices; they replied that if the Ohio Valley people would pack their pork and beef properly, the product would be in better demand. No doubt some had been packed incorrectly, but the basic source of difficulty was the use of Kanawha salt in Ohio Valley packing. This salt, described by one traveler as "tinged of a dirty red," was satisfactory for salting cattle, but the lime and magnesia in it rendered it unsuitable for packing, because these elements attack instead of preserve meat. Thus, many ship operators at New Orleans refused to carry Ohio Valley beef or pork for a sea voyage, and nobody would dare to ship it to the West Indies. Beef packed in the North Atlantic states commonly sold at New Orleans for twice as much as the competing Ohio Valley product. It was reported in 1821 that every barrel of Ohio Valley beef or pork had to be repacked with Turks Island salt when inspected at New Orleans, at a cost of $1.00 per barrel; and even when so repacked, it was inferior by $2.00 to $3.00 per barrel to provisions originally put up with proper salt. In vain did the Ohio Valley people argue that Onondaigua salt from New York was just as bad as Kanawha because of the artifical introduction of lime to remove the iron.[61]

The Ohio Valley cattleman found himself pinched. The Kanawha salt lessened the earning power of his barreled beef at New Orleans, yet the association of Kanawha saltmakers was engaging in price fixing.[62] The packer passed the cost of the salt on to the cattleman. The Kanawha salt manufacturers believed they were only protecting themselves from cutthroat competition, which presently, however, appeared from outside and uncontrollable sources; with the opening of the great salines of southern Illinois and of Missouri, the Kanawha salt industry entered a long and unsuccessful struggle for survival.

[61] David Thomas, *Travels through the Western Country in the Summer of 1816* (New York, 1819), 133; *Western Star*, December 8, 1821; Berry, *Western Prices*, 218; *Franklin Farmer* (Frankfort, Kentucky), January 11, 1840.
[62] They posted their price in the Chillicothe newspaper. See *Scioto Gazette and Fredonian Chronicle* (Chillicothe, Ohio), issues of 1819.

STOCKYARDS AND SLAUGHTERHOUSES 157

By October 29, 1821, Ohio Valley beef was selling for $6.50 to $8.50 a barrel on the New Orleans levee, but things were looking up. The increasing use of steamboats was effecting a drastic reduction of upstream freight rates; and as a result, natural sea salt was presently admitted to the interior at a low price. At the same time, the Kanawha brand began to undergo a series of improvements in quality and reductions in price.[63]

Even when improved, pickled beef could not compare with fresh beef, which the people of New Orleans relished for their own consumption, although in that climate it had to be rushed from shambles to table. Accordingly, it was occasionally profitable to take or send live cattle to that market. After 1842 the competition offered by Texas Longhorns, driven overland to New Orleans, was serious.[64]

The Ohio Valley cattlemen, moreover, had trouble with steamboats. In fact, Nat Hart had more trouble loading cattle on the steamboat at Louisville than Tom Shelby had in unloading at Pittsburgh: "On Sunday evening we put 31 of the last lot on board the Edward Strippen with much labour and noise and confusion, the sailors and Negroes doing it in their own way and not listening to me at all, and getting both lots first and last all or nearly so in the River, and some of them nearly drowned when we got them out. Of the last lot we left 5 that we could not get on board untill they became so furious from bad management, that we had to leave them standing in the pond."[65]

There were more things wrong with sending cattle on steamboats than just the fact that some beasts were agitated by getting on board, riding the steamboat, and getting off. Ordinarily, through shipment could be booked without too much difficulty from such major towns as Cincinnati and Louisville to New Orleans, and for such long-distance shipments on a trunk stream the rates were relatively low. But that was only the beginning. If the steamboat were not a regularly scheduled

[63] *Western Star*, December 1, 1821; Berry, *Western Prices*, 218.
[64] Towne and Wentworth, *Cattle and Men*, 154-55.
[65] Nathaniel Hart to Virginia Shelby, February 13, 1840, Filson Club.

packet, and most were not, departure might be delayed for several days while freight dribbled in. The feed the cattle would consume at the river town during this time might be expensive. Once under way, the skipper of a transient steamboat might linger for several days at some downriver town, in hope of picking up a choice lot of freight; hence it was best for the cattleman, if shipping on a transient, to have the freight charge include feed, and then if the cattle were emaciated upon arrival in New Orleans, he would know that he had been cheated on feed. The steamboat rate to New Orleans from Louisville in 1853 was $10.00 per head, presumably including feed.[66]

The cattleman could go with the cattle on the steamboat (as Nat Hart's son William did in 1840), and that suggested the old practice of taking them yourself, or sending a hired man with them, on a flatboat. This method avoided the disadvantages of steamboat shipment, and even counting the time of the man running the flat, it was perhaps a little cheaper; besides the usual risks of navigation, there was only the danger of getting "shoved" by a steamboat. Accordingly, flatboating lasted well into the heyday of the steamboat. Grade Durhams were being flatboated from Kentucky and Illinois to New Orleans in the spring of 1842.[67]

Beef receipts at New Orleans showed a general upward trend from 1825 until about 1850. In 1825 the receipts were 1,242 barrels; in 1828, 5,622; in 1831, 10,696; in 1834, 5,455; in 1837, 9,859; in 1840, 10,843; in 1844, 49,363; in 1848, 64,080 barrels and tierces. The years 1831 and 1835 (each of which is a high point in receipts with about 10,000 barrels) coincide with an improvement in beef prices, and 1831 also with the increased Ohio production of beef.[68] One can guess (but not prove) that in 1832, Ohioans, noting the New Orleans receipts of the year

[66] Hunter, *Steamboats*, 317-20; Richard Laverne Troutman, "Stock Raising in the Antebellum Bluegrass," *Register of the Kentucky Historical Society* (January, 1957), 55:19.
[67] *Dollar Farmer* (Louisville, Kentucky), March, 1843.
[68] James Hall, *Statistics of the West, at the Close of the Year 1836* (Cincinnati, 1836), 276.

before and regarding the New Orleans market as oversupplied, shifted their export eastward on hoof, producing the glut in the New York market from 1832 to 1834.

The drought in the Ohio Valley in 1839 and 1840 did not lessen beef receipts at New Orleans one iota—which suggests that the difference was being made up by Missouri beef. The agricultural depression that began in 1839 seems to have increased, at first, the beef receipts at New Orleans; some 33,000 barrels came down in 1841, and some 49,000 in 1844, in the pit of the beef depression. Of course, it might be said that the increase in receipts was, in turn, increasing the depression. The Whig tariff of 1842 raised the prices of some imports at the same time beef prices in New Orleans, New York, and Boston were falling in late summer of 1842; however, beef prices made a small rally in 1843. In 1845-1846, beef receipts at New Orleans at least held their own despite the fact that the Ohio River was closed by ice for a long period early in 1845. Because of the tightening of credit and the increase of Gulf freight rates, produce was piling up on the New Orleans levee. In 1848 receipts of some 64,000 barrels and tierces were the biggest of any recorded year up to then in the history of New Orleans.[69]

New Orleans was a major shipper (in bottoms owned by persons of other ports) of barreled beef to foreign countries. Usually this export nearly equaled or exceeded the total of New Orleans' shipments to United States ports. Of the barreled beef transshipped to United States ports, more went to New York and Boston than to any other coastal point; only occasionally did the sum total going to American ports other than Boston and New York exceed that going to Boston. New Orleans' beef export fluctuated sharply. In 1844, when 49,000 barrels were received from the interior, less than 5,000 were exported; thus, even after the home needs of New Orleans were filled, thousands of barrels of beef were accumulating on the waterfront because of the depression. In 1847, with prices much better, New York and Boston together took about 18,000

[69] *Hunt's*, June, 1842; June, 1843; June, 1846; October, 1848; October, 1849.

barrels, and foreign countries about 33,000. The next year, although prices receded, a large part of the flood of beef arriving at New Orleans was exported; about 20,000 barrels went to New York and Boston together, and about 33,000 abroad.

Monthly exports from New Orleans depended on the condition of the rivers, down which most of the barreled stuff had come. In the banner year of 1848, receipts were heaviest in January, next highest in March and December.[70] Apparently the winter traffic carried the product of November slaughtering and utilized the high water ordinarily accompanying winter, in spite of the risk of ice.

Before the railroad era, Ohio Valley drovers and cattlemen called at Boston only rarely. The first Ohio cattle (so far as we know) were driven in 1842 by William Renick, who found upon arrival that this market already had acquaintance with Kentucky cattle. Other Ohio Valley cattle were driven north from New York City by buyers from Boston. John T. Alexander's herd of Illinois cattle in 1845 brought $31 per head delivered here. Alexander had cattle here again in 1855; this drove was taken off the railroad cars at Bergen, and after a week's "holding back" there, was driven on foot to Boston, where they were sold for less than he had been offered in Illinois. By this time, with railroads having reached the Ohio Valley, cattle from Ohio, Indiana, and Illinois were much in evidence in the Boston market; many of these had been brought up from New York City by their owners, or the owners' agents, or Boston buyers.[71]

Boston, being the tanning center of the United States, took all the hides that the local butchers could provide, and probably many of those of the New York butchers as well.[72] The

[70] Sinclair, *The Port of New Orleans*, 175; *Hunt's*, October, 1844; October, 1845; October, 1847; September, October, 1848.

[71] W. Renick, *Memoirs*, 58; Thomas Shelby to M. C. Shelby, May 9, 1847; McClelland, "Strawn and Alexander," 202, 203; *Ohio Farmer*, June 30, 1855; May 3, 1856.

[72] In fact, the manufacture of boots, shoes, and leather had the largest gross of any industry in Massachusetts in 1845. *Tabular Representation of the Present Condition of Boston* (Boston, 1851), 22.

arrival of Ohio Valley cattle in numbers in Boston in the early 1850's followed closely the decline of California as a source of hides.

Cincinnati, centrally located to the cattle regions of the Ohio Valley, might be supposed to have been a more important cattle market than it was. Actually, in all its recorded history up to 1846, Cincinnati's consumption was 183,416 cattle; its export, 194,570. This is not impressive when contrasted to Philadelphia receipts, which totaled more than 190,000 cattle for just the four years 1844-1847.[73] The pork trade of Cincinnati—truly the hog butcher of the Ohio Valley—was of course much more voluminous than the beef trade; both hogs and cattle were slaughtered in establishments north of the canal and along Mill Creek, a stream that ran red with blood.

In the fall of 1821, a year of extreme depression, Andrews and Shays, originally a Boston firm, erected on Front Street "extensive warehouses for packing beef, pork, and lard in the best manner." It turned out that in their second year of business in the new buildings on Front Street, Andrews and Shays packed only 100 barrels and 163 half-barrels of beef, but 2,394 barrels of pork, 339 hogsheads of hams and shoulders, 3,406 kegs of lard, and 17,355 pounds of bacon. Some of the explanation may be the Southerners' taste for pork, the slowness of the beef-cattle industry to develop in Cincinnati's tributary Miami Valley, and the fact that farmers could not get much if any more for beef than for pork. Nevertheless, in the early 1820's the number of cattle slaughtered was sufficient to supply hides to a thriving leather-manufacturing industry, an industry which, in fact, paced most of the city's early enterprises.[74]

Cincinnati, as already mentioned, became a kind of preserve for the cattle feeders of the Kentucky Bluegrass and the Miami

[73] Charles Cist, *Cincinnati Miscellany* (Cincinnati, 1845-1846), 157; *Niles' Register*, February 5, 1848.

[74] Berry, *Western Prices*, 219, 220; *Lebanon Farmer*, January 24, 1817; *Western Star*, March 28, 1820; Cincinnati Federal Writers of the Works Progress Administration, *They Built a City: 150 Years of Industrial Cincinnati* (Cincinnati, 1938), 163.

Valley. As early as 1822 Warren County was supplying high-quality beef to this city. Apparently Cincinnati as well as Pittsburgh was a chief market for the fat cattle of the upper Miami Valley, for many droves of cattle were observed on the "Montgomery Pike" bound for Cincinnati. By the late 1840's Cincinnati had become a market for Clinton County fat cattle.[75] One Cincinnati newspaper thought in April, 1843, that it would make sense for the Scioto Valley cattlemen to take some of their cattle to Cincinnati, where, because of undersupply, they would get (the paper claimed) as good a price as in the East and save the expense of transmontane droving. Although that April the prices at Cincinnati were not as good as New York, some other times they were as good or better; the trouble was that the Cincinnati market, though often undersupplied, was still limited in contrast to the great market concentration of New York, Philadelphia, and Baltimore.[76]

The Cincinnati market was usually characterized by lower prices than New York—a natural condition. Thus, in 1836 some Cincinnati butchers themselves took cattle to New York City, and a Kentucky observer on Manhattan believed they would make money on them. Although in July of 1847 the Cincinnati and New York markets were approximately the same, by August 13 the Cincinnati prices ($1.00-$5.00 per 100 pounds) were substantially lower than those in the eastern markets at the same time. On January 3, 1849, word reached the upper Miami Valley that twenty head of cattle just sold at Cincinnati had brought only $4.35. This was even lower than New York prices when western cattle started arriving there "in a bunch" in the spring, and the New York prices that year were below the national average. The Cincinnati cattle prices in the first week of April of each year, 1848 through 1853, were lower than the New York prices in every instance, though following the New

[75] Jones, "The Beef Cattle Industry," 194; *Western Star*, April 6, 1822; Josiah Morrow Scrapbook, in possession of the Philosophical and Historical Society of Ohio, Cincinnati; Ohio State Board of Agriculture, *Fourth Annual Report* . . . 1849, p. 70.

[76] *Cincinnati Chronicle*, cited in *Scioto Gazette*, April 27, 1843, Aldrich Report, 2:24-25.

York prices upward; but the average of prices at Cincinnati for the full month of April of each of these years was higher than New York, apparently because Ohio Valley cattle began crowding the New York market later in the month.[77]

By the 1840's Cincinnati, like Louisville, had some butchers who did not seek to buy cattle, but instead packed beef for a fee. One such was William Neff, who also had a cattle farm near Cincinnati and had made an importation of breeding stock from Britain. Neff advised his customers what prices the two byproducts of slaughtering, hides and tallow, might be expected to bring in Cincinnati.[78]

The transshipping of barreled beef at Cincinnati was insignificant, being far outranked by pork and especially whisky. Most of these provisions to be transshipped at Cincinnati arrived on the Miami Canal. In 1844 Cincinnati imported 1,203 barrels and 556 tierces of beef; in 1848, 348 barrels and 27 tierces.[79]

Cattle and beef remained relatively minor factors in the Cincinnati trade, although after 1846 Cincinnati's export of live cattle steadily increased. The export of packed beef in 1844 was 14,476 barrels and 3,552 tierces; in 1845 it was 17,000 to 20,000 barrels and tierces. Next year, Cincinnati consumed 31,200 beef cattle in twelve months; one-third of these were packed for the foreign market, and the export that year was 19,000 barrels and tierces. For the years 1851-1855 the average annual export was 32,000 barrels and tierces; for the years 1856-1860, 23,000 barrels and tierces. Just how much of this packed beef went to New Orleans is unknown; if three-fourths of it went there, it would have constituted about one-fourth of New Orleans' receipts of packed beef in the 1840's. Meanwhile, Cincinnati exported on the average about 500 live cattle

[77] Abner Cunningham to Brutus J. Clay, July 31, 1836; Aldrich Report, 2:24-25; *Scioto Gazette*, August 18, 1847; *Xenia Torch-Light* (Xenia, Ohio), January 4, 1849; *Price Current* (Cincinnati), quoted in *Ohio Cultivator*, November 15, 1853.

[78] *Ohio Cultivator*, August 1, 1845; William Neff to Thomas H. Shelby, November 20, 1845, in University of Kentucky Library.

[79] *Hunt's*, October, 1849; Commissioner of Patents, *Report, 1847*, p. 616.

per year from 1846 to 1850; about 5,400 per year from 1851 to 1855; and about 20,600 per year from 1856 to 1860.[80]

By 1836 there were two large packing establishments at St. Louis, one at Alton, and several up the Illinois River. St. Louis got its first stockyards in 1845, but it had long been a livestock market.[81]

The fact that St. Louis was a western city did not necessarily mean that cattle prices there would be unattractive. As occasionally happened on or near the frontier, there was a shortage of provisions in Missouri Territory in the fall of 1818, because of the recent heavy influx of settlers; beef reached $6.00 per hundredweight at St. Louis on November 9. Again in 1849 and the several years following, the California Gold Rush created an unusual demand for beef cattle; buyers snatched up all cattle arriving at St. Louis.[82]

There is no evidence, however, that Kentucky or Ohio cattle ever entered the St. Louis market. This was an obvious market for Illinois; receipts from Illinois must have been especially heavy during the late 1840's, a brief period when the cattle barons of Illinois had already begun large operations but the railroads had not yet arrived from the East. At times one of the Illinoisans, Jacob Strawn, was able to control the St. Louis market.[83]

This city shipped and transshipped both live cattle and barreled beef. Some 6,000 live cattle were sent from St. Louis down the river to New Orleans in 1841, and the number increased in subsequent years. Quincy, on the Illinois shore upriver from St. Louis, slaughtered 900 beef cattle the same year. Some 478 head of cattle and 4,000 barrels of beef were unloaded from boats at St. Louis in 1844. Ice in the river down-

---

[80] *Niles' Register*, February 7, 1846; Cist, *Cincinnati Miscellany*, 139-40; Berry, *Western Prices*, 222; *Hunt's*, October, 1849.

[81] J. M. Peck, *New Guide for Emigrants to the West* (Boston, 1836), 291; Clifford D. Carpenter, "The Early Cattle Industry in Missouri," *Missouri Historical Review* (April, 1953), 47:210.

[82] *Western Star*, January 11, 1819; Carpenter, "The Early Cattle Industry," 203.

[83] Helen M. Cavanagh, *Funk of Funk's Grove* (Bloomington, Illinois, 1952), 58.

stream from St. Louis discouraged packing at the beginning of 1845, but receipts for that year turned out to be up: 522 live cattle and 5,000 barrels of beef. The following winter (1845-1846) an estimated 6,000 head of cattle were slaughtered at St. Louis. The 1846 import was 12,346 barrels; the export, 20,113. By 1848 the import was up to some 17,000 barrels and tierces; the export, 38,000. Then, in the 1850's, the railroads dealt a blow to St. Louis' function as a river shipper and transshipper; only 5,400 barrels of beef came to the St. Louis waterfront in 1853 (and of these only 755 barrels from Illinois).[84]

Pittsburgh was an early market for Ohio Valley cattle. In 1814 Zadok Cramer listed beef, at 6 or 7 cents per pound, first in his nonalphabetical list of market quotations. The Bayardstown slaughterhouses were in operation by the 1840's. Pittsburgh paid for western beef by sending iron, glass, and cotton manufactures west. The city reexported some of the beef to Baltimore or Philadelphia for home consumption or exportation. Most of the cattle fattened around Springfield, Ohio, about 1840 were marketed at Pittsburgh.[85] Mention has already been made of Tom Shelby's bringing cattle by steamboat from Maysville to Pittsburgh in 1848.

Although no cattle from north of the Ohio River ever (so far as we know) came to Charleston, South Carolina, this market was not an exclusive preserve of the Kentucky cattlemen; Charleston also received cattle from Tennessee, North Carolina, and Virginia. Walter Chenault, the Kirkersville, Kentucky, cattleman, had a partner who lived in Charleston and helped him market cattle here. On October 23, 1846, the partner wrote Chenault that this market might be crowded with North Carolina cattle until Christmas. James Gay was here at the beginning of 1848 as agent for Strauder Goff and apparently some other Clark County cattlemen. Some of the

[84] *Hunt's*, August, 1846; April, 1854; Carlson, *The Illinois Military Tract*, 89; *Niles' Register*, February 7, 1846; Commissioner of Patents, *Report, 1847*, pp. 595-96; *1848*, p. 807.
[85] Cramer, *The Navigator* (1814), 67; *Hunt's*, April, 1844; Oliver S. Kelly, "Springfield as Remembered Sixty Years Ago," *Yester Year in Clark County* (Springfield, Ohio, 1947), 1:30.

Kentucky cattle sold here were contracted for before they arrived. In the spring of 1848 Lillard extended his contract with the butchers, though with, as Gay put it, "the express understanding if his stock came in and was not full fat the butchers were to make such deductions as they thought proper." Gay was worried that the butchers would introduce into the market competition from Tennessee, North Carolina, and Virginia; and well he might have worried, for if half the Southern cattle being marketed at New York City during these years had gone to Charleston instead, there would have been a frightful glut.[86]

Louisville, in contrast to Charleston, was, as we have mentioned, an exclusive preserve of the Kentucky cattlemen. Here the chief butchers in the 1830's were two Germans, Peter Klepenroof and Fred. Bremaker. They contracted for cattle in advance, and Bremaker in December of 1836 was "in some of the upper counties looking for cattle." In 1841, however, droves from the Inner Bluegrass were apparently started out for Louisville without contract. The Louisville cattle market, like most except possibly New York, was always highest in winter. While some cattle were sold by prearranged contract, and others in the open market, still others were not sold at all but were packed for a fee by Thomas Maxcy. The fee in 1848 was 50 cents for killing and dressing each beef and $1.50 per barrel for packing (including the charge for the barrels.) Like Neff at Cincinnati, Maxcy advised the cattlemen what they might expect to get for hides and tallow.[87]

Chicago, boomtown of the Jacksonian era, arose as a cattle market at about the same time that a large-scale cattle industry arose in Indiana's Grand Prairie and in Illinois. No Ohio or Kentucky cattle (so far as we know) came to Chicago before 1860, although of course that development was just in the

[86] Walter Chenault to Brutus J. Clay, October 23, 1846; James Gay to Strauder Goff, January 1, April 28, 1848.
[87] William P. Hart to Virginia Shelby, December 3, 1836, November 6, 1841; Thomas Maxcy to Thomas H. Shelby, September 11, 1845, Maxcy to William F. Bullock, September 11, 1845, in University of Kentucky Library.

offing. The first frame building devoted to packing was erected by George W. Dole, located at what is now South Water and Dearborn Streets, in 1832. That year he packed 150 to 200 Indiana cattle. In the next three years, several slaughterhouses appeared, and others were established later. Cattleman Isaac Funk's brother, Absalom, was one of the Chicago butchers.[88] Chicago was a great market for grass cattle, which, depending upon their age and condition, were either butchered or sold as stockers to cattle feeders. Isaac Funk and Jonathan Cheney were reported many years later to have procured the first drove of stockers ever purchased in Chicago and brought to McLean County for feeding. Later, Funk appeared again in Chicago, this time about January 1, 1849, with a drove of 1,200 grass-feds which he had bought as stockers and was now marketing; he got nearly $30,000 for the drove, which averaged 677 pounds per head. The beef was "cured and packed in Mr. Dyer's best method," and sent east. Chicago got its first stockyards in 1848, when the Bull's Head yards opened at the corner of Ashland and Madison, the latter street being the route of produce wagons and drovers. The first yards accessible to railroads were opened about 1856, on the lake shore north of 31st Street.[89]

The volume of Chicago's beef export during the 1840's was apparently very inconsistent: 762 barrels in 1842, 10,000 in 1843, none in 1845, and 19,000 in 1848.[90] Not until the 1850's was Chicago a major exporter of beef. Most of the cattle then were coming from Illinois, Indiana, Wisconsin, Iowa, Missouri, and Texas.

Thus, the marketing pattern that emerged was just as complex as the grazing and feeding pattern. Even the eleven market cities discussed do not comprise a complete list; a few Ohio Valley cattle, for example, went to Richmond, Virginia, Nashville, and Cleveland. Such long-distance marketing of cattle

[88] Towne and Wentworth, *Cattle and Men*, 310; Clemen, *Meat Industry*, 83-84, 104-105; Cavanagh, *Funk*, 46-49.
[89] Cavanagh, *Funk*, 51; *Chicago Democrat*, cited by *Ohio Cultivator*, January 15, 1849; Clemen, *Meat Industry*, 83-84.
[90] *Hunt's*, February, 1848; November, 1849.

by the cattlemen, in person or through salesmen, has now, of course, become almost a thing of the past in the Ohio Valley. Most farmers now truck their cattle to the stockyards at the county seat, where they negotiate with the buyers from the packinghouses. This practice of the butchers' agents coming to the farmers had its origin, as we have noted, at least as early as the 1830's, when agents of Baltimore and Louisville slaughterhouses came to the Inner Bluegrass looking for cattle; but then it was an uncommon exception to long-distance marketing. Today, most of the county seats in the cattle-fattening regions are on railroads, and the buyers send the cattle in cattle cars or in motortrucks to the packinghouses. Most of the packing industry, in turn, has left the eastern seaboard, where the population is densest, and has gone west through the Corn Belt to meet the cattle—a development made possible by refrigerator cars, which take fresh beef to the consumers on the eastern seaboard.

Chapter 7

The Industry Moves Westward

BY THE MIDCENTURY THE CORN BELT WAS TAKING SHAPE. THIS became, not long after 1860, a geographical region of staggering immensity—250,000 square miles stretching from Columbus, Ohio, almost to the northeast corner of Colorado. The decisive factor in its emergence in Ohio and Indiana was conquest of the drainage problem; in Illinois, drainage plus railroads. Once the Corn Belt began to develop, its settlers drew upon the knowledge and the breeding stock of the older feeding regions. If the Corn Palace at Mitchell had a hall of fame for the founders of the Corn Belt, names like Renick, Selsor, and Funk should be in it.

Ohio Valley cattlemen had long been interested in the trans-Mississippi country that was to become the western part of the Corn Belt. Scarcely had the herds been established in the shadow of Mount Logan and Pilot Knob before the cattlemen were thinking about and visiting the country beyond the Mississippi—Missouri's Salt River bottoms, the Boon's Lick country, and the 'Tit Saw Plains. Their interest in Missouri was threefold: as a possibly superior cattle country in which they might settle; as a possible source of thin cattle that would be cheaper than those available in the Ohio Valley; and, eventually, as a possible market for breeding stock.

Perhaps because of the difficulties involved in expanding the cattle industry into Indiana and Illinois, the cattlemen had already investigated, by 1820, the first and second of these Missouri possibilities. Missouri became a cattle range at about the same time as Illinois. Travelers in Ohio in 1816 were astonished, upon encountering herds of range cattle clogging the trails and new roads, to learn from the drovers that the cattle had been trailed from Missouri Territory.[1] The animals were said to be on their way to the eastern markets, where rangey beef would still have buyers because the Pennsylvania and Kentucky feeders had not filled up the market, and because the Ohio feeders had appeared only twice (so far as we know) in the eastern market.[2] On their way through Ohio, however, some of the thin cattle from Missouri must surely have been taken up by Ohio feeders, who during the next ten or fifteen years made at least one drive to the eastern markets every year or two.

About 1825 Green Clay sent his sons Brutus and Sidney to explore Missouri; but Ohio cattlemen had already preceded them. In the spring of 1819, when the "Missouri fever" was at its height, Felix Renick and his younger brother William (who also had been fattening cattle the preceding winter) had decided to ride out and look the country over. As they explored the territory on horseback, as far north as a point opposite Quincy, Illinois, and as far west as a point about forty miles from Independence, they made field notes, which have survived. Though they surveyed the land with the thought of buying, they found that "the [Mississippi] Bottoms are extensive and very rich, but a great proportion of them lies too low and are subject to annual inundations which causes a great part of them to be too wet for cultivation and they must inevitably be sickly ... the disadvantages ... induces us to leave it without making any entries to try to find something more to our minds."

Striking southwest on horseback from the Salt River country, they encountered insufferable swarms of prairie flies. This scourge brought them to an abrupt halt. "We strick camp and after takeing some refreshment which we had brought with us we held a counsel" at which it was decided to travel single file by night to escape the flies. Eventually they hit the Franklin Road and reached the Boon's Lick country, where many of the creeks, they noted, "are ... sufficiently brackish for stock," being fed in large measure by salt springs. This, like the Salt

[1] David Thomas, *Travels through the Western Country in the Summer of 1816* (Auburn, New York, 1819), 120.
[2] George Renick's drive of 1805 and Felix's drive of 1815.

River country they had left behind, was already becoming a seat of the cattle industry in Missouri because of its salt, soils, and grasses; but the Renicks probed farther west—the Blackwater Fork of the Lamine (which after the Civil War was to become a great cattle country), the 'Tit Saw Plains, and the Sugartree Bottom (part of which was in the Missouri Military Bounty Tract, and which extended to a point about forty miles from Independence). William, although he may have had some bounty scrip, decided to make entry not in the Sugartree Bottom but on the 'Tit Saw Plains, where he became one of the first handful of settlers; but Felix preferred his Indian Creek farm back in Ohio.[3]

While thus, by 1820, stock cattle were going from Missouri to the Ohio Valley, and cattlemen were emigrating from the Ohio Valley to Missouri, the remaining possibility, that of Missouri's being a market for Ohio Valley breeding stock, received apparently little attention before 1820. There was not enough good quality breeding stock to go around in Ohio or Kentucky, let alone enough to supply Illinois or Missouri. Good quality breeding stock in 1815-1820 meant just that—good quality but not necessarily pureblood. The herds of Felix Renick at this time were an example; every well-formed bull in his herd had plenty of Ohioans clamoring to have him serve their cows. Pureblood stock was even rarer, there having been but two Shorthorn importations direct from Britain to the Ohio Valley by this time, those of Sanders and Prentiss, in 1817 and 1818, respectively. The initiative, therefore, for the movement of breeding stock to Missouri would have had to come from the Missourians, and there is no documentary evidence of it this early. The Missourians needed breeding stock, but lacked the resources to obtain it. Nevertheless, the assumption would be that some Missourians had obtained some breeding stock from

[3] *Western Star* (Lebanon, Ohio), January 26, 1819; "Journal of F. and W. Renick on an Exploring Tour to the Mississippi and Missouri Rivers in the Year 1819," and "Memorandum of Travel of F. and W. Renick and George Davies," in possession of Renick Cunningham, Chillicothe, Ohio. These manuscripts were published in *Agricultural History* (October, 1956), 30:174-86.

their relatives in the Ohio Valley by 1820, and that the scrubs the Missouri settlers had brought with them or purchased from the French were beginning to undergo slight upbreeding.

The movement of Ohio cattle to Missouri can be documented as early as 1826. In August of that year Felix Renick sold some of his cattle to George Davies, who had settled on the 'Tit Saw Plains. The fact that brother William was frequently in debt to Felix suggests that he, too, may have had some of Felix's cattle.[4]

A company or syndicate was formed in Ross County in 1832 to bring stock cattle from Missouri—a forerunner of the Ohio Company established the next year to bring blooded cattle from England to Ohio for breeding. The company appointed Felix and Thomas S. Renick as agents to go to Missouri to assemble a herd. The two men separated in Missouri and then met for the drive back. They purchased from one to forty-two head at a time, from forty-four different Missourians. The prices varied from $9.25 a head to $21.00; many were $10.00. Throughout the half century, the traditional price for stockers in Ohio was $10.00 to $12.00; therefore, the fact that the Renicks were willing to buy Missouri stockers at this price suggests that there was an extraordinary demand for stockers that year in Ohio, attributable perhaps to the deep penetration Ohio feeders were making in the eastern markets after 1831. In other words, demand for and supply of stockers were momentarily out of balance.

The drive back was much like hundreds of others into or out of the Ohio Valley, but it is the only one of which, to my knowledge, a trail journal has survived. In St. Louis, the Renicks must have excited quite as much curiosity as they did the next year in England with their "procession stirred up by a few Yankees." At the head of the herd rode Felix Renick, cattle king of Ohio, with a rifle and a "grizley bear skin" on the saddle in front of him, a tall black hat such as the Amish wear on his

[4] Felix Renick Bank and Memo Book, Renick Cunningham collection.

head, and an umbrella held over him to keep the sun off. Behind came the lowing herd of 220 Missouri cattle of assorted quality. Of these, two perished on the remainder of the drive. The Renicks' total expenses going to Missouri and coming back with the drove (feed, ferriage, etc.) were $381.09, of which $153.55¼ was chargeable to the firm, the rest being personal expenditures.

The drive reached the Scioto Valley in July. Two hundred of these Missouri cattle were sold to the Ohio feeders by February, 1833, bringing $4,905. The company's profit, therefore, was about $2,150. Three hides were sold and one steer butchered and sold by the company.[5]

The Renicks' drive was not the only one. About the same year, 3,000 cattle in one year crossed the Mississippi at the Clarksville, Missouri, ferry on their way to Ohio or Pennsylvania.[6] This did not even include many smaller droves going to Illinois' Sangamon Valley.

The position of Missouri as a kind of hinterland to the Ohio Valley cattle industry galled many Missourians, one of whom, Nathaniel Leonard, decided about 1838 to do something about it. A Vermonter who had settled in Cooper County about twelve miles from Boonville, Leonard had left New England, but had not left all his kin far behind; a sister had married a Kentuckian, James S. Hutchinson, and his brother Ben lived in Chillicothe, Ohio. Desirous of bringing to Missouri not only good, but also pedigreed, breeding stock, Leonard formed a partnership with his Kentucky brother-in-law and wrote to his brother Ben about the subject.[7]

Ben agreed to look around Ross County and see what was

[5] The account of the company and of the cattle drive it sponsored is based on Felix Renick's Memo Book, Acts of 1832 and 1833, Renick Cunningham collection.

[6] Clifford D. Carpenter, "The Early Cattle Industry in Missouri," *Missouri Historical Review* (April, 1953), 47:201.

[7] John Ashton, *History of Shorthorns in Missouri prior to the Civil War* (Jefferson City, Missouri, 1923), in Missouri State Board of Agriculture, *Monthly Bulletin* No. 11, 21:13, 14-16. Ashton's account is based on manuscript material discovered at the Leonard farm in Missouri.

available. Felix Renick, he found, had a pedigreed bull for sale, but Ben thought it was too leggy. George Renick, however, offered for sale a pedigreed white bull which Ben considered to be of better conformation—and it could be bought for $600 because it had had a spell of losing weight. This was Comet Star—among the purest of purebreds, for its sire was Comet Halley himself, and its dam too was an animal of the Ohio Company's importation. Ben closed the deal for Comet Star and at the same time purchased one of George Renick's heifers for $500. The two animals were loaded onto a canal boat at Chillicothe's Water Street to begin their journey to Missouri. Such great importance was attached to the pedigrees that these papers were taken in person to Missouri by an armed custodian, Thomas Boyen. Comet Star is believed to have been the first pedigreed Shorthorn bull to enter Missouri.[8]

Nathaniel Leonard's insistence on pedigrees did not make sense to his brother, who complained that it excluded from consideration many fine breeding cattle. Ben was undoubtedly influenced in this opinion by his admiration for the unpedigreed bull Accomodation, which Walter Angus Dunn had just brought from Kentucky to Ross County. Accordingly, Ben Leonard on his own purchased for his brother two bulls and one heifer, all of Kentucky stock, described somewhat enigmatically by Ashton as "Kentucky full-blooded but not thoroughbred stock." Anyway, the animals looked good to Ben, and in his opinion they were good buys—in fact, he liked one of the bulls better than Comet Star."[9]

The decade following the arrival of Comet Star in Missouri was marked by a growth of both the range and feeding branches of Missouri's cattle industry, until in 1850 that state had about seven beef cattle for every free male engaged in agriculture. During 1841, 6,000 live cattle, mostly from Missouri, were sent from St. Louis on steamboats down the river to New Orleans. By 1845 the cattle trade of St. Louis, which

---
[8] Ashton, 16-21; Alvin H. Sanders, *Short-Horn Cattle* (Chicago, 1900), 347.
[9] Ashton, *History of Shorthorns*, 16-21.

was dominated by Missouri and Illinois, had reached such proportions that a stockyard was built. Missouri did her share in boosting the receipts of both live and packed beef at New Orleans to levels during the 1840's four to seven times as high as the levels of the 1830's, despite the fact that more than half of the decade of the forties was a time of agricultural depression. The survival of the range in Missouri is evidenced by the poor quality of many of the Misouri cattle observed on the steamboats going to New Orleans.[10] The depression of the forties cut off most importations of pedigreed Shorthorns from England at a time when Missouri had made no direct importations and had imported only a few pedigreed Shorthorns from the Ohio Valley.

To the cattlemen of the 'Tit Saw Plains the army contract had always been a help, and with the acquisition of Oregon in 1846 the army market increased. The California Gold Rush sent demand soaring above the heretofore plentiful supply, so that cattle buyers from Independence and St. Joseph went into eastern Missouri looking for cattle, and even into Illinois and Kentucky, to the Paris cattle market.[11]

The end of the agricultural depression, the demand for beef in California, and the nationwide "bull market" for beef in the early 1850's inspired some cattlemen all the way from Independence to the Scioto to bring in Texas or Indian cattle. Later, after the idea had been tried out and cattle prices were climaxing in the crazy market of 1855, Solon Robinson again urged the plan: "Our suggestion of buying cattle in Texas and shipping them here [New York City] is still a good one. With a capital or credit of $100,000, any smart drover can pocket $50,000 profit in the first sixty days. It is what the sporting men

[10] U. S. Bureau of the Census, *Statistical View of the United States, being a Compendium of the Seventh Census* (Washington, 1854), 128, 170; *Hunt's Merchants' Magazine and Commercial Review* (New York), June, 1843; October, 1845; August, 1846; November, 1848; October, 1849; Carpenter, "The Early Cattle Industry," 210; James Hall, *Statistics of the West, at the Close of the Year 1836* (Cincinnati, 1836), 276; Herbert A. Kellar (ed.), *Solon Robinson, Pioneer and Agriculturist* (Indianapolis, 1936), 2:124.
[11] Carpenter, "The Early Cattle Industry," 203.

would call 'a sure thing.' Who will have the money?"[12] Plenty would, and a few of them were Ohio Valley cattlemen. In their minds this project was to be a moneymaking sideline; few had any intention of letting a Texas bull get within pole's length of his Shorthorn cows. Although the Renicks found some fairly good Texas Shorthorn stock which had been taken to the Southwest by settlers from the Ohio Valley and Missouri, the typical cattle in Texas were Spanish Longhorns, a breed later made famous by the cowboys of the Great Plains.

The first recorded cattle drive from Texas to the Ohio Valley had occurred, apparently, in 1846. Edward Piper herded these Longhorns to Ohio, where presumably they were fattened for market.[13] From 1849 on, drives from Texas to Missouri became fairly common. The Texas cattle were sold at Kansas City, Independence, and, in the late 1850's, St. Louis. Some of the cattle were then driven on to eastern markets, crossing the Mississippi at Quincy, Hannibal, and towns farther south.

In the summer of 1852 an Illinoisan, Joseph Mallory of Piatt County, ventured into the Cherokee nation (in what is now Oklahoma) and bought up cattle from the Indians. Mallory trailed the cattle to Illinois, where he fed them for almost a year. His idea was to send them to the booming New York cattle market; but when the herd reached the railhead at Laporte, near Michigan City, Indiana, an Ohioan of the west-central Ohio upland approached Mallory and bought the entire drove. The Ohioan, Seymour G. Renick, was a cattleman himself, but he had no intention of taking these Cherokee Longhorns to his farm; this was to be pure speculation, for he sent the cattle directly on their way from Laporte to New York City. Their arrival at the New York stockyards excited much curiosity, a newspaper reporter observing, "They all bear their original owner's brand, some of the figures of which belong to the Cherokee alphabet." They were surprisingly well received,

[12] *New York Daily Tribune*, May 10, 1855.
[13] Clarence P. McClelland, "Jacob Strawn and John T. Alexander, Central Illinois Stockmen," *Journal of the Illinois State Historical Society* (June, 1941), 34:200.

THE INDUSTRY MOVES WESTWARD 177

for the butchers took them at $65.00 to $78.00 per head—which probably meant about $8.00 per 100 pounds.[14]

Others were encouraged. The following year, 1853, Tom Candy Ponting, an Illinoisan, assembled a herd of Texas Longhorns. These he trailed north through the Indian Territory, paying tolls to the various chiefs along the way. Passing through the Missouri border country apparently without incident, the drive proceeded through Springfield, Missouri, to St. Louis, where the herd crossed into Illinois by ferryboat. The cattle were wintered somewhere in Illinois and reached the New York market on July 3, 1854.[15]

Cherokee Longhorns are noted again in Illinois in 1854. On July 16, a hundred head of these cattle, "of the large, broad horned order," were driven through Springfield, Illinois, on their way to the prairie pasturelands northwest of Bloomington. Thirteen days later another herd—a huge one of 1,200 head—clogged the dusty streets of the Illinois capital. It was William Renick's herd of Texas cattle, which subsequently became well publicized. These cattle, Renick explained, had been purchased in northern Texas, "which section of the country had been to a considerable extent settled by immigrants from Illinois and Missouri, and who had brought their stock with them; and this stock had not yet been sufficiently intermixed with the Spanish or Opelousas cattle further south to materially deteriorate their original qualities; consequently they were a much better and larger stock than I expected to see, though they had in some measure acquired the wild nature of the more southern stock."

The Ohioan, who during the 1830's had become contemptuous of Kentucky stock not only as breeders but also as market beef, appears surprisingly tolerant of Texas stock. Upon their arrival in Springfield they were, he claimed, "in fine order and well broke." The people of Springfield, all of whom got a good look at them, thought otherwise. He was able to sell few or

---

[14] Joseph G. McCoy, *Historic Sketches of the Cattle Trade of the West and Southwest*, ed. by Ralph P. Bieber (Glendale, California, 1940), 40-41.
[15] Carpenter, "The Early Cattle Industry," 204.

none of them to Springfield butchers or to Sangamon feeders, but it is known that he sold some in Chicago during the same year and others in New York City during the spring of 1855.[16]

While William Renick was trying to find buyers in Illinois for his Texas cattle, George Squires of Illinois was down near Houston dickering for a herd of five hundred Longhorns to bring back. The deal having been closed, the herd was assembled at Austin, where another Illinoisan, John Bent, joined Squires to help with the drive. Squires' wife, who had come along south for a holiday in New Orleans, met the men at Austin. Finally the outfit got under way, the two Illinoisans on horseback, with perhaps several cowboys, and Mrs. Squires riding in a covered wagon at the head of the herd. Like Ponting, they struck directly into the Indian Territory, but thereafter they took a more northerly route, traversing a corner of Kansas. At the historic cattle crossing of Hannibal, they swam the entire herd across the Mississippi. It is unknown how many if any of the cattle were lost between Austin and Illinois. The herd was pastured near Bent's farm on the open prairie a few miles south of DeKalb, or else near the present site of Riverside, Illinois.[17]

The Renick and Squires drives, and the drive of Cherokee Longhorns, were not the only ones to cross the Mississippi into Illinois during 1854-1855. On September 11, 1854, six hundred Texas cattle were offered for sale at Douglasville, Illinois, opposite Hannibal. About the same number reached the Chicago stockyards during the year. A man named Taylor of Missouri trailed 154 Longhorns from Texas to Illinois and sold them to William Rankin, who wintered them in Illinois and sent them to New York City in the spring of 1855. Isaac Funk, onetime Ohioan and later one of the Illinois cattle barons, is said to have purchased in 1855 with John Nichols 1,200 head of Texas

---

[16] *Illinois Daily Journal* (Springfield), July 17, 1854; *New York Daily Tribune*, July 4, 1854; William Renick, *Memoirs, Correspondence, and Reminiscenses* (Circleville, Ohio, 1880), 24; McCoy, *Sketches of the Cattle Trade*, 38.

[17] George S. Herrington, "An Early Cattle Drive from Texas to Illinois," *Southwestern Historical Quarterly* (October, 1951), 55:267-69.

cattle for $27,000. To Illinois, too, throughout the fifties came a dribble of cattle from Arkansas, Kansas, Nebraska, and Iowa —all types of scrubs and some sleek Longhorns that did not sell any better than the conventional stockers of the Illinois prairie.[18]

The reputation of the plains cattle was indeed spotty. The first Longhorns in the Chicago market fetched about as low a price as Ohio Valley cattlemen had even known, even in bad times—$2.50 to $2.75 per 100 pounds. Seymour Renick's Cherokee Longhorns, however, proved more rewarding. At least twenty-one of the Ponting-Malone herd brought $80.00 each in the New York market, and since they had been fed for only one year, this seems to mean at least $9.00 to $9.50 per 100 pounds. Solon Robinson, at the New York stockyards at the time, remarked that the meat of Texas or Indian Territory Longhorns could be improved "by purchasing them young and feeding them two years as well as this drove has been fed one year."[19]

Damaging to the reputation of the Texas Longhorns was the fact that some of them had Texas fever. In 1855 this fever infected many of the Shorthorn herds of Missouri, and the local farmers endeavored—with but scant success until late 1858—to keep Texas cattle out of Missouri. Occasionally they formed an armed force to stop the cowboys at the county lines.[20]

But some cowboys still reached New York City; Longhorns kept arriving that had been on the move every day the whole route from Texas to New York. If the New York housewife balked at this beef, the general beef boom, which occasionally narrowed the price differential between good and bad cattle, allowed it to be salted and packed for export—a process that made tender beef tough anyway. Some of the plains beef,

[18] McCoy, *Sketches of the Cattle Trade*, 38; Helen M. Cavanagh, *Funk of Funk's Grove* (Bloomington, Illinois, 1952), 51-52; *New York Daily Tribune*, June 14, 1855.
[19] McCoy, *Sketches of the Cattle Trade*, 39, 41; *New York Daily Tribune*, July 4, 1854.
[20] Carpenter, "The Early Cattle Industry," 204.

however, had undergone improvement, such as Robinson suggested, by longer stopovers in the Ohio Valley for fattening, and hence it was acceptable as fresh beef. The Ohio Valley stopover was feasible if the Longhorns were free of Texas fever. During the late 1850's, the Illinois graziers, some of whom specialized in plains cattle, swamped the New York market with fat cattle both of plains origin and of local origin, which they sent to New York by railroad, forcing the Scioto feeders to the wall. The resultant leveling off of the New York beef boom in 1856-1857, with a slight dip in 1858, must have caused the slaughterers to reject, or at least to discount heavily in price, plains Longhorns that had not been "tenderized" in the Ohio Valley. The Illinois graziers met the challenge by doing more feeding, and the movement of plains Longhorns to New York remained fairly constant through 1859. About 750 arrived in 1855, about 1,000 to 2,000 per year from 1856 through 1859.[21]

The railroads connected Illinois not only with the cattle markets, but also, in the late 1860's, with the plains range. Joseph G. McCoy was one of those who saw the opportunity, for Texas had a surplus of cattle, but the North, a shortage. McCoy, who was to become the builder of Abilene, had grown up in the Sangamon cattle country of Illinois, had got his start by shipping a carload of mules to the famous mule market of Paris, Kentucky, and in 1861 decided to become a cattle trader.[22]

The trade in plains cattle in the Ohio Valley before 1860 occurred entirely on the north side of the Ohio River. Central Kentucky did not see Texas Longhorns until 1868, when Joseph Scott brought about 500 of them to Paris to trade for mules. Other Kentuckians thereupon discovered that they could buy Texas Longhorns as stockers cheaper than they could buy the stockers available in the Ohio Valley, and they proceeded to act on this discovery sporadically for about ten or fifteen years

[21] McCoy, *Sketches of the Cattle Trade*, 42; Robert Leslie Jones, "The Beef Cattle Industry in Ohio prior to the Civil War," *Ohio Historical Quarterly* (June, 1955), 64:317-18.
[22] McCoy, *Sketches of the Cattle Trade*, 42.

—to the detriment, according to one authority, of Kentucky's beef-cattle industry.[23]

While plains Longhorns were moving east, Ohio Valley breeding stock were moving west across the Mississippi in small but significant numbers, so that by 1860 about seventeen Missourians had registered Shorthorns. When J. Stoddard Johnston began cotton farming in Phillips County, Arkansas, in 1855, he brought cattle with him from the Kentucky Bluegrass, where he had been a cattle trader. Brutus Clay sold breeding stock during the late 1850's to farmers around Mexico in central Missouri. Missourians purchased some of the Shorthorns at the Warfield sale in 1856. The Shakers of Pleasant Hill, Kentucky, sold thirty Shorthorns in or near St. Louis in 1857. And James B. Clay, Henry Clay's eldest son, of Fayette County, Kentucky, offered a whole herd of Durham cattle for auction during the St. Louis Fair in 1859.[24]

The Ohio Valley cattlemen had been interested in Missouri as a place to settle, as a source of stockers, and as a market for breeding stock. The quest for feeding cattle eventually led to the importation of Texas or Indian Territory Longhorns. It was in Missouri, eastern Kansas, and eastern Nebraska that the spreading cattle industries of Texas and the Ohio Valley met. There they swirled about each other, the Ohio Valley industry contributing breeding stock, feeding techniques, and actual working experience along with the grazing and feeding pattern. The Ohio Valley industry was not entirely strange to the Texas cattlemen; many of them, particularly those around Dallas and Fort Worth, had migrated to Texas from the Ohio Valley, where they had had their first introduction to the lore of cattle

[23] Elizabeth Ritter Clotfelter, "The Agricultural History of Bourbon County, Kentucky, prior to 1900" (thesis, University of Kentucky, 1953), 20.

[24] Sanders, *Short-Horn Cattle*, 348; J. Stoddard Johnston, "Farm Book and Journal of events more especially bearing on Agriculture" (1855-1865), in possession of the Filson Club, Louisville, Kentucky; Brutus J. Clay, Register of Names and Memorandum Book, 1854-1878, in possession of Cassius M. Clay, Paris, Kentucky; *Valley Farmer* (Louisville), January, June, 1857; James B. Clay, *Catalogue of Durham Short Horns the Property of James B. Clay . . . which will be sold . . . the week of the St. Louis Fair* (Lexington, Kentucky, 1859).

raising and where they had had their first experience with the "long drive." After the defeat of the Sioux in 1876, the rich grasslands between the Sand Hills and the Bozeman Trail, and those in southern and western Montana, became available, and the cattlemen migrated with their stock from Missouri, Kansas, and Nebraska. Following them into the area came thousands of Oregon and Washington cattle, some of them third- and fourth-generation descendants of Ohio Valley cattle, veterans of the Oregon Trail.

By the late 1880's Illinois could no longer compete with Iowa and northern Missouri. Eastern Iowa first offered competition; then southwestern Iowa, around Glenwood, Red Oak, and Shenandoah. Across the Missouri border, a great cattle empire grew around Tarkio and in the Nodaway Valley, the northwestern limit of settlement by Southerners, and around Maryville. That was in the northern Missouri prairie, whose development by steam power Felix Renick had prophesied in 1819.[25] The center of fat-beef production moved westward from central Illinois, leaving the Ohio Valley forever.

The ante bellum cattlemen of the Ohio Valley were, in a sense, farmers engaged in the toil of making a living. This toil plus vision—the ability and willingness to form big plans—wrought accomplishments: a system of successfully grazing, fattening, and marketing cattle; improved breeds of cattle; and the introduction of this experience, as well as good breeding stock, to the lands west of the Mississippi. If we note the relatively short span of time involved and the initial problem of occupying the Ohio Valley, we find the list of accomplishments impressive. So did the contemporaries of these men; agriculture then enjoyed a prestige that has not been equaled since. The cattlemen built a distinctive way of life which has survived, to some extent, wherever men raise cattle.

[25] Towne and Wentworth, *Cattle and Men,* 226; Renick Journal.

Bibliographical Note

THE MOST IMPORTANT MATERIALS FOR A STUDY OF THE BEEF-cattle industry in the Ohio Valley before 1860 have never been published. These include the pioneer work in the field, Charles Townsend Leavitt, "The Meat and Dairy Livestock Industry, 1819-60" (dissertation, University of Chicago, 1931), of which only small portions have been published from time to time in *Agricultural History*. The other very important unpublished materials are the Brutus J. Clay papers, in the possession of Cassius M. Clay, Paris, Kentucky, which include a wide correspondence with cattlemen all over the Ohio Valley; and the Shelby family papers in the manuscript room of the library of the University of Kentucky. Only slightly less important are the Strauder Goff correspondence, in the possession of Ben Douglas Goff, Sr., Winchester, Kentucky; the Felix Renick papers, in the possession of Renick Cunningham, Chillicothe, Ohio; and the Hart family papers, in the possession of the Filson Club, Louisville, Kentucky. Leavitt used none of these manuscript sources.

The most valuable published work for the general reader is probably James Westfall Thompson, *History of Livestock Raising in the United States, 1607-1860*, United States Department of Agriculture, Agricultural History series, No. 5 (Washington, 1942). Also helpful as a recent, specialized study is Robert Leslie Jones, "The Beef Cattle Industry in Ohio prior to the Civil War," *Ohio Historical Quarterly* (April, June, 1955), 64:168, 315; Jones, however, made inadequate use of the *Scioto Gazette* and ignored almost all manuscript sources relating to Ohio and to the Ohio-Kentucky trade. The beef-cattle industry in the Grand Prairie of Indiana is adequately treated in Paul Gates, "Hoosier Cattle Kings," *Indiana Magazine of History* (March, 1948), 44:1. The breeding story has never really found its way into print; what we have, instead, are highly technical catalogs, of which the outstanding one is the authoritative Alvin H. Sanders, *Shorthorn Cattle* (Chicago, 1900), now a somewhat rare book.

The agricultural press is a source of prime importance, notably the *Franklin Farmer* (Frankfort and Lexington, Kentucky), the *Kentucky Farmer* (Frankfort), the *Ohio Cultivator* (Columbus), the *Ohio Farmer* (Cleveland), and the *Western Farmer* (Cincinnati). But a combing of county-seat newspapers also proves rewarding, especially the *Western Citizen* (Paris, Kentucky) and the *Scioto Gazette* (Chillicothe, Ohio). The latter has been preserved from 1800 to the present with gaps of only a very few years, and the microfilms in the newspaper's office in Chillicothe should be an important source for historians of early agricultural history and the westward movement. The library of the University of Chicago has the *Western Citizen* and the *Maysville Eagle* (Maysville, Kentucky).

No small amount of material is found in government documents. The U. S. Agricultural census, of course, is a basic source. Accounts of crop and weather conditions may be found in the *Reports* of the U. S. Commissioner of Patents. The Ohio State Board of Agriculture's *Annual Reports* include county-by-county breakdowns of yields, etc.

Travelers' accounts offer a scattering of material about cattle. Especially useful, because he was earlier than the crowd of travelers, is David Thomas, *Travels through the Western Country in the Summer of 1816* (Auburn, New York, 1819). Guidebooks or statistics books of the time deserve mention: Daniel Drake, *Natural and Statistical View of Cincinnati and the Miami Country* (Cincinnati, 1815); James Hall, *Statistics of the West, at the Close of the Year 1836* (Cincinnati, 1836); and J. M. Peck, *New Guide for Emigrants to the West* (Boston, 1836).

Index

Accomodation: admired by B. Leonard, 174
Acmon: bought by M. L. Sullivant, 78
Adams County, Ohio: as early source of stockers, 15; as later source of stockers, 57
Adelaide family: origin, 80
Akron, Ohio: on drover route, 109
Alexander, John T.: as cattleman, 65; helped drive cattle in late 1830's, 117; made drive in 1846, 117; in Boston market in 1845 and 1855, 160; mentioned, 89
Alexander, Robert Aitcheson: made importations during 1840's, 89; shipped Sebastopol to Kentucky, 91; bought Sirius, 92; as breeder, 92-93; at Lexington fair in 1855, 93, 100; imported Duke of Airdrie, 93; deplored prejudice for Bates Shorthorns, 95
Allen, John and Albert: at Lexington fair, 100
Allen, Lewis F.: compiled *American Herd Book*, 23; on Patton cattle, 26; disagreed with A. Waddle, 33n; assembled pedigree for Mrs. Motte, 34
Alton, Illinois: packing at, 164
American Herd Book, 23
Amesville, Ohio: mentioned, 104
Anderson, William M.: mentioned, 34
Andrews and Shays: Cincinnati packers, 161
Apples: as slop for cattle, 68
Asheville, North Carolina: cattle passing through in 1840, 117
Ashton, John: mentioned, 174
Athens County, Ohio: exported stockers, 57
Atkinson, Thomas: herded cattle in Benton County, 18; made drive to Pennsylvania, 116
Auvergne: B. Clay estate, mentioned, 44
Ayrshire cattle: recommended to J. Dunn, 75; came into Ohio, 87

Bagg, J.: mentioned, 44
Baker, Henry: traded with E. Harness and J. Van Meter, 51; mentioned, 50
Baltimore: butchers sent agents to Kentucky, 113; as cattle market, 152-54
Bank of Chillicothe notes, 10
Banks: lent money, 10
Banton, W. T.: mentioned, 7
Barnesville, Ohio: mentioned, 122
"Barren Cattle" of west-central Ohio: described, 17
Barren County, Kentucky: feeding entered, 56-57
Barrens of southern Kentucky, 3. *See also* Green River Barrens
Barrington, Hon. Miss: given to wife of Episcopalian bishop, 77
Bates Shorthorns: described, 91; prejudice for, 94-95
Bedford, George: answered by L. Sanders, 30-31; as part owner of Sebastopol, 91; secured guarantee of Locomotive's pedigree, 91; owned Bell Duke of Airdrie, 96; at Lexington Fair of 1855, 100; mentioned, 44
Bedford, Pennsylvania: on drover route, 108; condition of roads near, 124
Bedford Springs, Pennsylvania: on drover route, 108
Bell Duke of Airdrie: owned by G. M. Bedford, 96
Belvidere, Illinois: mentioned, 139
Benton County, Indiana: as part of early range, 18; as surviving range, 62
Bermuda: cattle exported to, 150
Big Hill Gap: on Boone trace, 1
Big Levels: received B. Clay stock, 92
Big Levels of West Virginia: as cattle country, 2
Blackwater Fork of the Lamine: mentioned, 171
Blanton, H.: put cattle "to keep" at Scott's, 48

INDEX

Bluegrass: characteristics of, 5; origin in North America, 5n; in Scioto Valley, 9; in Illinois, 19n; mentioned, 43

Bluegrass Basin of Kentucky: early cattle grazing and feeding in, 5-8; Ohioans got stockers from, 14; later cattle grazing and feeding in, 42-48

Blue Licks, Kentucky: visited by Felix Renick, 14

Boone, Daniel: viewed Bluegrass Basin, 1

Boon's Lick Country of Missouri, 169, 170

Boston, Massachusetts: as cattle market, 160-61

Bourbon County, Kentucky: as source of stockers, 14, 15; as beef county, 42; later feeding in, 44; Durhams grazed in Fayette County, Ohio, 61; as hemp land, 70; got Otley, 80; received animals of Fayette Company importation, 82

Boyle County, Kentucky: as part of Lincoln County, 42, 43

Brahman cattle: reached Ohio Valley, 40

Bremaker, Fred: Louisville butcher, 166

Brent, Charles S.: brought Powell cattle to Kentucky, 39

"Broadlands" or Sullivant farm: acquired by J. Alexander, 65

Brown, James N.: owned Lady Macallister, 83; grazed and fed cattle, 64; moved to Illinois, 88; bought Rachel, 95, 100-101

Brown, J. P.: mentioned, 50

Brown County, Ohio: as early source of stockers, 15; as later source of stockers, 57

Brownsville, Pennsylvania: had bridge over Monongahela, 107

Brush Creek cattle: described, 15

Buffalo: crossed with cattle, 39-40

Buffenberger, Peter: "buck prairie" grazier, 57-58

Bundy, Ezekiel: drove stand, 122

Butler County, Ohio: as beef county, 53

Buzzard: brought to Kentucky, 27

Calif and Jacoby: contest with B. Harris, 65

Cambridge, Ohio: fork in road, 107

Campbell and Brown: Chillicothe packers, 132

Captain Balco, 98

Carlisle, Pennsylvania: on drover routes, 108

Carolina cowpens. See Cowpens, Carolina

Carroll County, Kentucky: mentioned, 105

"Centerville" whip, 119

Chambersburg, Pennsylvania: on cattle trail, 108

Champaign County, Illinois: as part of early range, 19; mentioned, 111

Champaign County, Ohio: prairies of, 15; early feeding in, 16

Charleston, South Carolina: as cattle market, 165

Charleston, West Virginia: on cattle route, 105

Cheapside Market, Lexington: stockers bartered at, 43

Chenault, Walter: mentioned, 46

Cheney, Jonathan: got stockers at Chicago 167

Chicago, Illinois: as destination of some drives, 110; as cattle market, 166-67

Chickasaw cattle: bought by Felix Renick, 14; blended into scrub breed, 21

Chillicothe, Ohio: Kentucky stock exhibited, 1835, 34; as stocker market, 48; on cattle route, 107; flatboats leaving for New Orleans in 1820, 155

Cincinnati, Ohio: received Warren County beef, 11; as cattle market, 161-64

Circleville, Ohio: strange cattle seen near, 32

Clark, C. M.: bought New Year's Day, 98

Clark County, Kentucky: as beef county, 42; later feeding in, 44-45; herds assembled in, 113; mentioned, 8, 105, 111

Clark County, Ohio: western, engaged in fattening, 53; export of

INDEX 187

Clark County, Ohio: *continued*
  cattle in 1850, 59; eastern, engaged in grazing, 61; mentioned, 116
Clark County (Ohio) Company, 98
Clarksburg, West Virginia: mentioned, 2
Clarkson, Charles S.: sold Durhams in 1838-1839, 81-82
Clarksville, Missouri: cattle crossing, 173
Clay, Brutus (son of Green): on Patton cattle, 26, 29; saw strange cattle in Scioto Valley, 32; asked for stock by Trumbull countian, 34; offered steers by T. Clay, 43; heard from J. Fox about stockers, 43; had slaves, 43; was left with Cassius' cattle in 1846, 46; made partnership with Cassius, 46-47; bought heavy stockers in 1854, 56; cattle noticed by Scott countian, 81; bought Matilda, 81, 83; built breeding herd, 83; traded stock with Dunn boys, 89-90; got letter urging Westfriesian breed, 90; as part owner of Sebastopol, 91; bought Diamond, 91; sold breeding stock to Illinoisans, Iowans, and Missourians 92, 101, 181; got report from R. Corwin, 97-98; suggested a sweepstakes tie at Springfield, 99; claimed vindication of Sander's stock in 1854, 99-100; mentioned, 44
Clay, Cassius Marcellus (son of Green): on Bluegrass, 6n; background, 45; as cattleman, 45-46; political and military activities, 46; bought Primrose I from J. Garrard, 83; mentioned, 70
Clay, "Graybeard" Sam: as cattle feeder, 45
Clay, Green: as cattle grazier, 7-8
Clay, Henry: imported Hereford cattle, 28; wrote letter of introduction for F. Renick, 76; recommended Herefords, 76; contributed cattle to drove in 1844, 117; mentioned, 7
Clay, Henry Jr.: urged turnips as cattle feed, 67; imported Shorthorns, 80
Clay, James B.: sold Durhams at St. Louis, 181
Clay, L. B. and Co.: exhibited Kentucky stock at Chillicothe, 34
Clay, Tacitus: offered steers for sale, 43
Cleopatra: bought by D. Sutton, 39
Clermont County, Ohio: later grazing and feeding in, 53
Clinton County Association, 98
Cloud, B. F.: as agent-drover, 8, 113; made drive in 1845, 117
Clover: in Illinois, 19n; introduced in the East, 22
Columbia, Pennsylvania: on drover routes, 108
Columbus, Ohio: on cattle route, 107, 110
Comet: imported, 80; bested by Comet Halley, 81; used in Ohio, 89-90
Comet Halley: sold by Ohio Company, 78; sired Pocahontas, 78; sired Rose of Sharon II, 79; on R. Pindell's farm, 81; bested Comet, 81
Comet Star: taken from Ohio to Missouri, 174
Corn: surplus by 1824, 4; species, 4, 71; surplus in Scioto Valley by 1803, 8; raising invaded west-central Ohio upland, 16, 60; crop size in relation to livestock pack, 41-42; good years, 41, 42, 49; sold to cattle feeders, 48; Scioto, Western Reserve cattle wintered on, 49-50; uses of in Scioto Valley in 1850's, 51; raising and feeding encroached on Illinois range, 63, 64-65; as "king" crop, 71; need for rotating, 71-72
Corwin, Robert: imported Scottish Shorthorns, 97; concerned about misrepresentations to Kentucky breeders, 97-98
Coshocton, Ohio: mentioned, 16
Cowboys: of early west-central Ohio upland, 16
Cowpens, Carolina: compared to west-central Ohio range, 16; cowboys of, compared to early herdsmen of west-central Ohio, 17
Crab Orchard, Kentucky: on Wilderness Road, 1
Cramer, Zadok: quoted beef, 165

Credit. *See* Banks
Crockett, Robert: mentioned, 7
Crop rotations, 54-55, 71-72
Crouse, John: sent drove to Philadelphia in 1854, 126
Cumberland Ford: cattle being driven through in 1822, 115-16; cattle being driven through in 1828, 116
Cumberland Gap: cattle being driven through in 1802, 103-104; cattle passing through in 1828, 105; cattle being driven through in 1821, 115
Cumberland Road or Pike. *See* National Road
Cunningham, Abner: made drive in 1836, 116; analyzed his mistakes in marketing cattle, 127; in New York market in 1836, 144
Cunningham, Isaac: bought half interest in Young Mary, 78; bought cattle at Powell sale, 81; mentioned, 28
Cutler, Ephraim: exported cattle eastward, 104; made drives, 114; mentioned, 130-31
Cynthiana, Kentucky: mentioned, 95
Cyrus: won certificate at Lexington Fair, 100

Dairying, 68, 73
Danville, Kentucky: stockers around, 43
Darby Creek: cattle feeding along, 60
Darby Plains: as pasture for Scioto cattle, 4
Darke County, Ohio: as range, 55
Davies, George: bought cattle from F. Renick in 1826, 172
Deer Creek: cattle feeding along, 50, 60
Devon cattle: quality of feed required, 13; brought by New Englanders, 21; discussed by Ohio Company, 76; later discussion and importations of, 86-87
Diamond: bought by B. Clay, 91; impotent, 91, 96
Dillard, R. T.: as agent for Fayette Company, 82
Distilling, 72
Dole, George W.: Chicago packer, 167
Douglasville, Illinois: mentioned, 178

Drenning: made drive in 1817, 115, 143
Drew, Daniel: made drive in 1818 or 1819, 115, 143
Drift, 111
Drought, 41, 49
Drover: types of, 112-13
Duchess (imported): won premier honors at U. S. Cattle Show, 99
Duchess (Rose of Sharon cow): exhibited at U. S. Cattle Show, 99
Dudley, Edwin: advocated eventual emancipation, 46
Dudley, Nelson: as agent for Fayette Company, 82; as agent for Northern Kentucky Company, 90
Duke of Airdrie: imported by R. A. Alexander, 93; rented by A. Renick, 93-94
Duke of Airdrie 2478, 94
Dunkirk, New York: on cattle trail, 109
Dunn, George: sold a drove in Philadelphia, 116
Dunn, John: journeyed to Britain in 1833, 75
Dunn, Walter Angus: land speculations in early west-central Ohio upland, 17, 17n; imported cattle in 1833, 75; imported Comet and four cows, 80; sold Durhams in 1838-1839, 81-82; sent Accomodation to Ohio, 89
Durham Cow, the: bred to San Martin and Napoleon, 30; bred to W. Smith's Longhorn bulls, 31
Durham Shorthorns: said to need good feed, 13; later discussion of, 83-84
Duroc: mentioned, 33

Eades, Thomas: advertised Brahman bull, 40
Earl. Adams: as grazier and feeder, 62; fed distillery slop to cattle, 72
Edwardsville, Illinois: range country around, 19
Egbert, Jacob: as cattle feeder, 52
Ellsworth, Henry L.: encouraged development of Grand Prairie, 61; exported cattle, 61; on prairie breaking, 61
Ely family: mentioned, 53

INDEX

English Herd Book, 23
Erie, Pennsylvania: on cattle trail, 109
Erie Canal: took some cattle traffic, 109
Exception: B. Clay's chief bull, 83
Exchange: mentioned, 33

Fairfield County, Ohio: mentioned, 110
Fayette County, Kentucky: as beef county, 42; later feeding in, 47; received animals of Fayette Company importation, 82
Fayette County, Ohio: as part of early range, 15; traveler's description of typical farmer of, in 1816, 16; summer-pastured Ross County cattle, 58; later grazing in, 58-59, 61; forest cleared, 60; got Ohio Company cattle, 77
Fayette County Importing Company, 82
Feeder cattle: defined 4n
Fencing: in west-central Ohio upland, 17, 59; on Illinois prairie, 62, 63-64
Finley, John: viewed Bluegrass Basin, 1
Fisher, Stephen: mentioned, 7
Flagg, G. W.: advertised Shorthorn bull 89
Flint, Timothy: described third wave of settlers, 10
Floods, flash: on Scioto and tributaries, 8, 8n
Flora Belle: sired by New Year's Day, 98
Foley, Jim: hired B. F. Harris, 116
Fort Schuyler, New York: on cattle trail, 109
Fowler, Capt. Jonathan: made drive in 1804, 114
Fowler, Moses: as grazier and feeder, 62
Fox, Joel: wrote B. Clay about stockers, 43
Franklin County, Kentucky: later feeding in, 47-48
Fraser's slaughterhouse: Chillicothe, 132
Funk, Absalom: Chicago butcher, 167
Funk, Isaac: sent herd to Buffalo, 116-17; in Chicago market, 167; bought Texas cattle in 1855, 178-79

Gallatin County, Kentucky: mentioned, 105
Garrard, Charles T.: as agent for Northern Kentucky Company, 90; as part owner of Sebastopol, 91; mentioned, 44
Garrard, Gen. James: liked the "Seventeens," 31
Garrard, James II: stock dispersed, 81; sold Primrose I to C. Clay, 83
Gay, James Douglas: on Patton cattle, 26; as Goff agent in the Charleston, South Carolina, market, 113, 165-66
Gay family: on South Branch, 2
Genesee Country: cattle being herded out by 1810, 104
Geneva, New York: on cattle trail, 109
Gettysburg, Pennsylvania: on drover route, 108
Goff, Strauder: as cattle feeder, 44-45; marketed cattle through an agent drover, 113; sent cattle to Charleston, South Carolina, in 1848, 117-18; received order from small-town butcher, 132; marketing cattle in Cincinnati, 133; sold cattle to agent of Baltimore butcher, 153; marketing cattle at Charleston, South Carolina, 165-66
Goff, Thomas: trip back to Virginia, 5; as cattle dealer, 8; made drive in 1816, 114; in Philadelphia market in 1816, 150; mentioned, 2
Goodwine family: drove cattle to Chicago, 126
"Gopher ditches," 61
Gough of Virginia: imported British cattle in 1783, 25
Grand Master: took premium at Lexington Fair, 100
Grand Prairie of Indiana: as early range or stocker country, 18-19; later grazing and feeding in, 61-62; mentioned, 117
Gray, Lewis C.: on Patton cattle as Devons, 26
Great Crossings, Kentucky: mentioned, 81
Great Genesee Road. *See* Ontario and Genesee Turnpike
Great Kanawha River: as pioneer route, 1

Green River Barrens: as part of range, 14; William Renick in, 15; grazing and feeding in, 57; mentioned, 66
Greenbrier River: mentioned, 2
Greene County, Ohio: land being cleared, 11; as cattle county, 53
Greensburg, Pennsylvania: on cattle trail, 108
Greenville, Fort, Treaty Line: mentioned, 3
Groom, Benjamin: as agent for R. A. Alexander, 93
Gwinn, Mr.: Ohio upland cattleman in 1832, 57

Hagerstown, Maryland: on National Road, 123
Hancock's Branch: mentioned, 8, 44, 45
Hannastown, Pennsylvania: on drover route, 109
Hannibal, Missouri: as cattle crossing, 176, 178
Harness, Edwin: as cattleman, 50; as Ohio Company agent in England, 76-77
Harness, Joseph: bought Young Mary, 78; sent a drove to New York in 1818, 115, 143
Harriet: designated for James Renick, 77
Harris, Benjamin "Frank": grazed and fed cattle, 64; contest with Calif and Jacoby, 65; as example of professional drover, 112; made drives in 1834-1835, 116; sent drove to Boston in 1854, 118; recounted attempts upon his moneybag, 124
Harris, Thomas, of Williamsburg, Virginia: engaged W. A. Dunn as agent, 17n
Harrisburg, Pennsylvania: on drover routes, 109
Harrison, Col. Beatal: dealt in cattle, 60
Harrison, Benjamin: disputed A. Sanders about Mars, 25-26
Harrison, D.: mentioned, 7
Harrison, William Henry: as Governor of Indiana Territory, 3; advocated crop rotations, 54-55, 72
Harrison County, Kentucky: as corn country, 42

Harrold family: as cattle feeders, 60
Hart, John: bought two bulls from Prentice, 39
Hart, Nathaniel: received land grant, 5; bought two bulls from Prentice, 39; switched from hemp raising to grazing, 70; advised his sister against shipping on flatboats, 134; had trouble loading cattle on steamboat, 157; mentioned, 47
Hart, William (son of Nathaniel): mentioned, 158
Hart, William P.: had difficulty finding drove assistants, 119-20; hired a hand to take cattle to New Orleans, 154
Heath, Lewis: as agent-drover, 8; assembled stockers for Felix Renick, 14
Hemp raising, 69-70, 73
Hendricks County, Indiana: as cattle county, 54
Henry County, Indiana: mentioned, 96
Herd Book: Dr. Samuel Martin's, 23; English, 23; Felix Renick's, 23; Kentucky, 23; L. Allen's *American*, 23
Hereford cattle: imported by Henry Clay, 28; discussed by Ohio Company, 76; later discussion and importations of, 86
Highbanks: Edwin Harness farm, 50
Highland County, Ohio: as early source of stockers, 15; some later feeding in, 57; southern, as part of stocker country, 57; got Ohio Company cattle, 77
Hitch family: mentioned, 53
Hocking cattle: described, 15
Hocking River: mentioned, 2
Hog raising: rivaled cattle industry in Ohio Valley, 72-73
Hollingsworth family: as cattle feeders, 52
Holmhurst: Goff estate, 44-45
Holt, Judge: mentioned, 53
Huffnagle and McCollister: Chillicothe butchers, 132
Hughes, Jacob: political, financial, agricultural activities, 47
Hughes, J. C.: agent for Baltimore butchers, 113; made drive in 1843, 117

INDEX

Hutchcraft, Reuben: imported seven Shorthorns in 1839, 82; mentioned, 44
Hutchcraft family: perhaps used Shaker, 28

Illinois: as source of cattle for Ohio graziers, 58
Illinois Importing Company, 100-101
Illinois prairie: as early range or stocker country, 19-20; as surviving range or stocker country, 62-65
Illustrious: used by Abram Renick, 77
Independence, Missouri: mentioned, 171, 175
Indian Creek: F. Renick farm, 27, 50
Indianapolis, Indiana: early cattle feeding northwest of, 11-12; reached by National Road, 12; on cattle route, 106, 110
Inskeep family: mentioned, 53
Irish cattle: reached America, 32
Irving, Peter: corresponded with Henry Clay, 28

Jackson's Purchase: settlement of, 3
James River and Kanawha Turnpike: as cattle trail, 105-106
Javelin: imported by J. S. Matson, 90
Jefferson, Ohio: grazing near, 61
Jessamine County, Kentucky: got one of Venus' calves, 25; received animals of Fayette Company importation, 82; mentioned, 85
John Bull: imported to Kentucky, 39
John O'Gaunt: imported by J. S. Matson, 90
Johnson, E., of Oldtown, Ohio: boasted of soil productivity, 16
Johnson, John: observed need for manuring, 54-55
Johnston, J. Stoddard: cattle trade, 85-86; took Kentucky cattle to Arkansas, 181
Joline: Chillicothe cattle merchant, 132
Josephine: consigned to Felix Renick, 77; used by Abram Renick, 77
Juniata River: as cattle trail, 108, 109

Kanawha salt: used in packing, 156-57
Kentucky Importing Company (Scott County Company), 92

Kentucky Stock Book, 23
Kerr, John: exhibited cattle in Cincinnati, 47
Kingwood, West Virginia: mentioned, 2
Kirk: taken to Montana in 1881, 101-102
Klepenroof, Peter: Louisville butcher, 166

Lady Durham: produce of, 33
Lady Macallister: mentioned, 33
Lady of Clark: sired by New Year's Day, 98
Lady of the Lake: bought by R. Seymour, 78; described, sold to G. Renick, 78; dropped calves, 78
Lafayette, Indiana: mentioned, 132
Lancaster, Ohio: on pioneer route, 1
Laura: exhibited at U. S. Cattle Show, 99
Laurel Hill, Pennsylvania: on drover route, 108
Leafland: Jacob Hugh estate, 47
Lebanon, Ohio: mentioned, 132
Lebanon, Pennsylvania: stall-feeding around, 103, 110
Leonard, Ben: selected stock for Nathaniel Leonard, 173-74
Leonard, Nathaniel: brought pedigreed stock to Missouri, 173-74
Lewisburg, West Virginia: mentioned, 2
Lexington, Kentucky: stockers bartered at, 43
Lexington (Kentucky) Fair of 1855, 100
Licking County, Ohio: received Skipton Bridge, 77; mentioned, 110
Ligonier, Pennsylvania: on drover route, 108
Limestone (Maysville), Kentucky: as landing place to Kentucky, 2
Lincoln, Levi: sent cattle to Alton, Illinois, 89
Lincoln County, Kentucky: as beef county, 42
Lindale, Ohio: mentioned, 53
Linson, George: mentioned, 58
Little Miami Railroad: trains killed cattle, 59
Locomotive: brought to Bourbon County, 91

Logan County, Ohio: mentioned, 15
London, Ohio: as cow town, 59; stock cattle market in, 59, 59n
Longhorn cattle: in the Sanders importation, 30-31; in Kentucky, 31, 34-37, 38; in Scioto Valley, 32
Lord John: took premium at Lexington Fair, 100
Lord Vane Tempest: stood at B. C. Bedford's farm, 91
Louan family of Shorthorns: origin, 39
Louisville, Kentucky: as cattle market, 166; mentioned, 95
Lutz, Samuel: made drive in 1822, 116

McArthur, Duncan: advertised Whitaker in Kentucky, 80
McConnell, James: had Devons, 89
McCoy, Joseph G.: became cattle trader in 1861, 180
McLean County, Illinois: as part of early range, 19; cattle fattening in, in 1850, 63; mentioned, 111, 116-17
McVeytown, Pennsylvania: stock farms around, 109
Mad River: cattle drive from, 114
Madison County, Kentucky: as beef county, 42; as source of stockers, 43
Madison County, Ohio: as part of range, 15; summer pastured Ross County cattle, 17, 58; later grazing in, 58-59; got Ohio Company animal, 77-78; mentioned, 118
Madison County (Ohio) Company, 97
Mallory, Joseph: drove Indian cattle to Illinois, 176
Manuring: in Scioto Valley, 54-55
Marietta, Ohio: mentioned, 3
Marshall County, Kentucky: as part of range, 14; shipped out fat beef by 1849, 57
Martin, Dr. Samuel: kept herd book, 23; as cattle feeder, 44; commented on manuring, 55; participated in 1839 importation, 82; theories on breeding, 84-85
Mary Ann family: origin, 80
Marygold: offspring of Oliver and Primrose I, 83
Mars: reached Kentucky, 25
Matilda: bought by B. Clay, 81, 83

Matson, James S.: advertised Durhams in 1850, 90; imported bulls in 1852, 90
Maxcy, Thomas: Louisville packer, 166
Medina County, Ohio: mentioned, 108
Mercer County, Kentucky: received animals of Fayette Company importation, 82
Mercer County, Ohio: as range, 55
Meredith, Gen. Sol: purchased Shorthorns, 89
Miami County, Ohio: mentioned, 53
Miami Importing Company: mentioned, 11
Miami Valley: early cattle feeding in, 10-11; got Patton cattle, 27; later feeding in, 52-53; slowness of beef cattle industry to develop in, 67n; cattle in Cincinnati market, 161-62, 165; mentioned, 2
Michaux, F. A.: mentioned, 103-104
Middletown, Virginia: feeding near, 115
Miller, of Maryland: imported British cattle in 1783, 25; sold bulls to Kentuckians, 27, 28
Miller, A.: bought Kentucky stockers, 14
Miller, Aaron: estimated graziers' profits, 60
Miller, John, 23
Miller, William, of Ohio: mentioned, 50
Miller cattle: descendants lauded and criticized, 28-29
Missouri: as source of cattle for Ohio graziers, 58
Mohawk River, as pioneer route, 2
Monocacy River: as cattle trail, 108
Montgomery County, Indiana: as cattle county, 54
Montgomery County, Kentucky: as part of range, 14
Montgomery County, Ohio: as beef county, 53
Moorefield, West Virginia: mentioned, 2
Morgan County, Illinois: mentioned, 116
Morristown, New Jersey: cattle tally at in 1849, 118
Morrow, Jeremiah: mentioned, 11

INDEX

Mount Lebanon: Garrard family estate, 44, 81
Mount Oval: William Renick home, 50-51
Mount Sterling, Ohio: William Renick bought cattle near, 18
Mount Vernon, Ohio: on drover route, 109
Mrs. Motte: bred to San Martin and Tecumseh, 30
Murrain, bloody: cattle disease, high losses from, 17-18
Murray, Bronson: got an imported Shorthorn, 89
Muskingum River: mentioned, 2

National Road: reached Indianapolis, 12; as cattle trail, 106-107
Neff, William: Cincinnati packer and breeder, 163
New Berlin, Illinois: prairie and farms near, 64
New Englanders: in Ohio Valley, 2; attitude toward breeds, 21-22, 23
New Orleans: as cattle market, 154-60
New Richmond, Ohio: mentioned, 53
New Year's Day: as show animal and as getter, 98; bested Perfection, 99
New York City: as cattle market, 142-50
Newton County, Indiana: as part of early range, 18
Nichols, John: bought Texas cattle, 178-79
Nigger Mountain: drove stand on, 122-23
"Nimblewill": a grass, in Illinois, 19n
Northern Kentucky Importing Company, 90-91
Northwestern Turnpike: as cattle trail, 106

Oakland Farm, Meredith Farm: mentioned, 89
Offut, A. D.: sold Colonel. 88
Offut, Otho: produced cattle, 47
Offut, Sam: mentioned, 44
Ohio Company for Importing English Cattle: organized, 75-76; discussed breeds, 76; first importation, 76-77; second importation, 77; crossed imported stock with local stock, 77; final importation, 77; first auction,

Ohio Company: *continued*
77-78; final auction, 78; significance of activities, 78-79; conclusion about auctions, 79; Kentucky press commented on sales, 79-80
Oldtown (Old Chillicothe), Ohio: mentioned, 16
Olean: mentioned, 115
Ontario and Genesee Turnpike: as cattle trail, 109
Orcharding, 68, 73
Otley: brought to Bourbon County, 80; sired calf by Matilda, 81
Owensboro, Kentucky: mentioned, 43

Paint Creek: corn land in Fayette County, 16, 60
Paint Creek, North Fork: mentioned, 58
Paragon of the West: given to F. Renick, 78
Paris, Kentucky: stock-cattle market in, 56; J. S. McCoy visited, 180; Texas cattle traded for mules, 180
Parkersburg, West Virginia: on cattle route, 106
Partridge family: mentioned, 53
Patton, James (son of Matthew): migrated to Kentucky, 25
Patton, John (son of Matthew): migrated to Kentucky, 25; moved to Ross County, Ohio, 27
Patton, Matthew: homestead, 8; migrated to Kentucky, 25
Patton cattle: defined, 26; lauded and criticized, 28-29; mentioned, 85
Patton family: on South Branch, 21, 25
Pennyrile: settlement of, 3; part of, as range, 14
Peoria, Illinois: cattle drive from in 1841, 117
Perfection: exhibited at U. S. Cattle Show 99
Philadelphia: as a cattle market, 150-52
Pickaway County, Ohio: exported cattle in 1819, 9, 115; western, as part of range, 15; later feeding in, 50, 51-52; got Ohio Company cattle, 77
Pierce, Col. Darlington: mentioned, 60

INDEX

Pierce, William: as cattle feeder, 60
Pike County, Ohio: later feeding in, 50-51; got Ohio Company animal, 77-78
Pilot Knob: mentioned, 1
Piper, Edward: drove Texas cattle to Ohio, 176
Pindell, Richard: had Comet Halley on his farm, 81
Pine Grove, Kentucky: mentioned, 44
Pittsburgh, Pennsylvania: rise of slaughterhouses, 108
Pleasant Hill Shakers: used Buzzard and Shaker, 27-28; sent trading boat down the Mississippi, 135-36; sold Shorthorns at St. Louis, 181
Pluto: brought to Kentucky, 27; taken to Ohio, 27; progeny of, 28
Pocahontas: born and bought, 78
Ponting, Tom Candy: drove Texas cattle to Illinois, 177; Texas cattle in New York market, 179
Potomac River, South Branch. *See* South Branch
Powell, Col. John H.: imported Durhams, 39; stock dispersed, 81
Powell County, Kentucky: as part of range, 14
Prairie, Illinois. *See* Illinois prairie
Pratt, John: noticed B. Clay's cattle, 81
Prentice, James: imported breeding cattle, 39
Prices, cattle, 1834-1860, 41
Primrose I: acquired by B. Clay, 83
Prince Regent: imported to Kentucky, 39
Putnam County, Indiana: as cattle county, 54; mentioned, 96
Putnam County, New York: herds being driven out by 1814, 104
Pynchon, William and Sons: meat packers, 130

Quincy, Illinois: as cattle crossing, 176; mentioned, 132

Rachel: bought by J. Brown, 95, 100-101
Range: early Kentucky, 14-15; later Kentucky, 56-57
Ranges: early, 12-20; later, 55-66

Rankin, William: wintered Texas cattle, 178
Reading, Pennsylvania: on drover route, 108
Renick, Abram: used Josephine and Illustrious, 77; used Harriet, 77; bought Lady of the Lake, 79; appointed to judge Ohio cattle, 80; rented Duke of Airdrie, 93; as breeder, 94-95, 96; sued by B. Van Meter, 95; mentioned, 45
Renick, Felix (1770-1848): cultural interests, 2; made drive to Philadelphia in 1815, 9; made drive from Missouri in 1832, 9, 10; as pioneer, 10; bought stockers from Chickasaws in 1810, 14; left money in Kentucky in 1815 for purchase of stockers, 14; went to Kentucky in 1816 to buy stockers, 14-15; kept herd book, 23; on Patton cattle, 26; bought John Patton's cattle, 27; bought "Patton" or other Miller-based cattle in 1816, 29; visited Strauder Goff, 44-45; died in accident in 1848, 50; suggested an importing company, 75; as Ohio Company agent in England, 76-77; received Josephine, 77; given Paragon of the West, 78; made drive in 1815, 114; in Philadelphia market in 1817, 150-51; explored Missouri in 1819, 170-71; sold breeding cattle into Missouri in 1826, 172; drove stockers from Missouri in 1832, 172-73; mentioned, 174
Renick, George, of Kentucky: on way to Kentucky, 1
Renick, George, of Ohio: brother of Felix, 8; fattened herd in 1804-1805, 8; estimated number of cattle fed in Scioto Valley, 9; estimated percentage of stockers locally raised, 12; bought Kentucky stockers, 14; bought John Patton's cattle, 27; bought Longhorn bull from W. Smith, 31; visited Strauder Goff, 44-45; retired and held auction 50; made drive to Baltimore in 1805, 103, 107; sent a drove to New York in 1818, 115, 143; made drive in 1833, 116; in Baltimore market

INDEX 195

Renick, George, of Ohio: *continued*
in 1805, 152; sold Comet Star to the Leonards, 174; mentioned, 131
Renick, George W., of Ohio (1796-1892) (son of Felix): as agent for Scioto Valley Company, 97
Renick, Harness (1810-1891) (son of George Renick of Ohio): as cattle feeder, 50; replied to B. Clay in 1855, 100
Renick, James (son of George of Kentucky, son-in-law of Felix): received Harriet, 77
Renick, Josiah: as Ohio Company agent in England, 76-77
Renick, Seymour G.: speculated in Indian cattle, 176; Indian cattle in New York market, 179
Renick, Thomas S.: drove stockers from Missouri in 1832, 172-73
Renick, William (d. 1845) (brother of Felix): explored Missouri in 1819, 170-71
Renick, William (son of George of Ohio): bought Missouri stockers in 1832 9; urged blooded cattle on rangeland farmers, 13; bought stockers in Kentucky, 1821-1825, 15; bought stockers in southern Ohio, 1821-1825, 15; lost cattle by murrain, 18; described some of his cattle, 34-35; cast implications upon Longhorns, 35; as cattle feeder, 50; made drive in 1824, 116; made drive in 1842, 117; on cost of beef, 137-38; commented on Philadelphia butchers, 151; drove Indian cattle to Illinois, 177-78; mentioned, 115
Renick family of Kentucky: in Greenbrier Valley, 2
Renick family of Ohio: on South Branch, 2
Riddling: defined, 6
Rising Sun (bull): on W. Smith's farm, 31
Robinson, Solon: comment on corn country, 12; on trading in Texas cattle, 175-76; on improving plains cattle, 179
Rose of Sharon I: killed, 79
Rose of Sharon II: sired by Comet Halley, 79

Rosedale, Ohio: mentioned, 17
Ross County, Ohio: early feeding in, 7-8; cattle from, summer-pastured in Madison County, 17, 58; agricultural society show, 1840, 34; topography contrasted to that of Kentucky Bluegrass, 42; later feeding in, 48-52; agricultural society show, 1835, 77
Rustling: in Illinois, 63

St. Joseph: mentioned, 175
St. Louis: as cattle market, 164, 165, 174-75
Salt family: mentioned, 53
Salt River Bottoms of Missouri, 169, 170-71
San Martin: imported, 30; as sire, 33; mentioned, 32
Sanders, Alvin: tells about Mars, 25; on Patton cattle, 26
Sanders, George N.: sold breeding cattle into Illinois, 88
Sanders, Lewis: imported British cattle in 1817, 29-30; defended his importation, 30-31; urged cross of Kentucky Shorthorns and European cattle, 84
Sanders family: on South Branch, 2
Sangamon County, Illinois: J. N. Brown moved to, 88
Sangamon Valley: early cattle feeding and grazing in, 12; later grazing and feeding in, 54, 62-65
Sawyer, Nat: sold his dairy cattle, 69
Scioto County, Ohio: got Ohio Company animal, 77-78
Scioto Valley: trade with South Branch, 4; early cattle feeding in, 8-10; strange cattle seen in, 32; later cattle feeding in, 48-52; manuring in, 55; mentioned, 1, 2, 92
Scioto Valley floods. *See* Floods, flash
Scioto Valley Importing Company, 97
Scott, Cunningham: appointed to judge Ohio cattle, 80
Scott, Robert W.: as cattleman, 47-48; manured his fields, 55; tried different varieties of corn, 71
Scott County, Kentucky: received animals of Fayette Company importation, 82

INDEX

Scott County Company (Kentucky Importing Company), 92
Scottish Bluebell: calved Flora Belle, 98
Scrub cattle: in stocker country, 13; advantages of, 21; attitudes toward, 21; strains in, 21
Sebastopol: shipped to Kentucky, 91
Selsor, David: went into grazing business, 57; bought corn, sold "Christmas beef," 60; traded stock with Dunn boys, 90; made drives east, 118
Senator: won certificate at Lexington Fair, 100
Settlement of Ohio Valley, 3
"Seventeens": defined, 30; generations, 32-33; controversy, 33, 35-39, 85
Seymour, Richard: bought Lady of the Lake, 79; made drive in 1824, 116
Shaker: brought to Kentucky, 27; progeny of, 28
Shakers. See Pleasant Hill Shakers; Union Village Shakers
Shelby, Gen. Isaac: made drive in 1842, 117
Shelby, Thomas I: received land grant, 5; made drive with J. Webb in 1845, 117; had bad time shipping cattle on steamboat, 134-35; attempted to reach the British market, 135; mentioned, 47, 145
Shelby County, Kentucky: as beef county, 42
Shirley, L. H.: sold Durhams in 1838-1839, 81-82
Shirley and Birch: bought cattle from S. Waite, 82
Shorthorn cattle: later discussion of, 83-84; fall from popularity, 102; mentioned, 23
Shotwell family: mentioned, 53
Shrink. See Drift
Shropshire, B. N.: bought Otley, 80
Sideling Hill, Pennsylvania: on drover route, 108
Sirius: bought by R. A. Alexander, 92; won premium at Lexington Fair, 100
Skipton Bridge: given to Episcopalian bishop, 77

Slaughtering: at Chillicothe in 1819, 9
Slaves: on cattle farms, 43; freed by Cassius Clay, 45; traded by Jacob Hughes, 47
Smith, Samuel: imported Comet and four cows, 80
Smith, William: spread Longhorn strain, 31; mentioned, 7
Soiling: defined, 6
Soils: of Ohio Valley, 1
Somerset, Pennsylvania: on drover route, 108
South Branch: as cattle country, 2; cattle trade, 4; got Scioto stockers, 8; continued trade with Scioto Valley, 51
South Carolina: Kentucky cattle on way to in 1835, 116
South Charleston, Ohio: as cow town, 16, 59-60; "buck prairie" near, 57; area, feeding entered, 60-61; near cattle routes, 106; drovers assembled cattle around, 113-14
South Mountain, Pennsylvania: on drover route, 108
Southerners: in Ohio Valley, 2; in early Illinois, 12
Spears, John: mentioned, 7
Springfield, Illinois: open range survived near, 63, 64
Springfield, Ohio: cattle fattening around, 53; as cow town, 59; as site of U. S. Cattle Show of 1854, 99; on cattle route, 106, 110
Squires, George: drove Texas cattle to Illinois, 178
Stall-feeding: description, 6-7
Staunton, Virginia: large droves passed through, 115
Steddom, Moses: as cattle feeder, 52
Steele family: mentioned, 53
Steubenville, Ohio: on drover route, 107-108; highwaymen, near, 123
Stevenson, Dr. A. C.: imported Shorthorns, 96-97
Stevenson, James D.: purchased Colonel, 88
Stock cattle: defined, 4n; imported to and exported from Scioto Valley in 1831, 9; market in Chillicothe, 48; problem of selecting, 55-56; market in Paris, 56; market in London, 59, 59n

Stock-cattle country. See Ranges
Stonehammer: taken by Ohio Shakers, 33
Store cattle: defined, 4n
Straw, wheat: as winter cattle feed, 60, 71
Strawn, Jacob: grazed and fed cattle, 64; and the St. Louis market, 64, 164
Strode's Creek: mentioned, 8
Sugar beets: as cattle feed, 67-68
Sugartree Bottom (Missouri), 171
Sullivant, M. L. and Co.: bought Acmon, 78
Sumner, Charles: sent droves to New York City in early 1850's, 118
Sumner, Edward: as grazier and feeder, 62; mentioned, 89
Sutton, David: bought Cleopatra, 39; stock dispersed, 81; Fayette Company held auction on his farm, 82
Switch cane, 5, 5n
Sylvia: produce of, 33
Syracuse, New York: on cattle route, 109

Taylor, Hubbard: participated in 1839 importation, 82
Taylor, J. P.: participated in 1839 importation, 82
Taylor family of Bourbon County: mentioned, 44
Tecumseh (bull): imported, 30; mentioned, 32
Teeswater Cow, the: offspring described, 30
Tennesseans: complained about Kentucky cattle, 37
Terre Haute, Indiana: on cattle route, 106
Texas cattle: driven to Ohio Valley and New York, 175-81
Timothy: in Illinois, 19n
Tippecanoe County, Indiana: as cattle county, 54, 62
'Tit Saw Plains of Missouri, 171, 175
Tobacco raising, 66, 69, 73
Toledo, Ohio: cattle arriving at by canal 11
"Trembles": cattle disease, 18
Trimble, Governor of Ohio: complained about Kentucky cattle, 37; mentioned, 57

Trumbull County, Ohio: mentioned, 34
Turnips: as cattle feed, 67-68

Union Village Shakers: sold some cattle in 1818, 11; used Shaker, 28; took Stonehammer, 33; as cattle feeders, 52-53; bought descendants of Whitehead importations, 88; got stock from Corwin importation, 97; bred Corwin stock, 98
United States Cattle Show of 1854, 99
Upland, west-central Ohio. See West-central Ohio upland
Utica, New York: on cattle route, 109

Van Meter, Benjamin: sued A. Renick, 95
Van Meter, Jacob: as cattle feeder, 50-51
Van Meter, Solomon: bought half interest in Young Mary, 78; as agent for Northern Kentucky Company, 90
Vanbrant and Adams: Baltimore butchers, 113
Vause, James Inskeep: as cattle feeder, 50-51; became a Democrat, 51
Vause, Joseph: as pioneer, 10
Venus: reached Kentucky, 25
Virginia (cow): as ancestress of Louan family, 39

Wabash River: flatboating cattle down, 11
Wabash Valley: early cattle feeding in, 11-12; later cattle feeding in, 53-54
Waddle, Alexander: questioned purity of Mrs. Motte, 33; as agent for Clark County Company, 98; mentioned, 97
Wadsworth family: cattle appeared in New York City in 1820's, 104, 143; used hired men and agents as drovers, 113; planned to get Ohio cattle in 1818, 115
Waite, Samuel: imported cattle through New Orleans, 82
Walke, John: mentioned, 50
Warfield, E.: mentioned, 7
Warfield, William: skeptical of pedigrees of Waite importation, 82

INDEX

Warren County, Kentucky: feeding entered, 56-57
Warren County, Ohio: later feeding in, 52-53
Warwick: bought cattle at Powell sale, 81
Washington Court House, Ohio: railroad opened to Wheeling, 126
Wasson: bought Otley, 80
Watts, Dr. Arthur: as cattle feeder, 50; as agent for Scioto Valley Company, 97; as agent for Clark County Company, 98; tied sweepstakes at Springfield, 99; approved of Sander's stock, 100
Webb, John: made drives in 1843-1847, 117; recounted troubles crossing the mountains, 124; in New York market in 1844, 145
Wellsville, Ohio: on drover route, 108
West, Richard: sold cattle to J. S. Johnston, 85-86
West family: mentioned, 53
West Newton, Pennsylvania: on drover route, 108
West-central Ohio upland: as early range or stocker country, 15-18; got yearling steers from Miami Valley, 52, 58; as later range or stocker country, 57-61; development as a breeding center, 97
Western Reserve of Ohio: wintered cattle in Scioto Valley in 1845-1846, 49-50; dairymen got cow calves from Miami Valley, 52
Wheat raising, 71
Wheeler, Bronson: Louisville commission man, 155
Wheeling, West Virginia: on pioneer route, 1; highwaymen near, 123

Whig tariff of 1842, 49, 49n
Whisky. *See* Distilling
Whitaker: advertised in Kentucky, 80
White Hall: post office, 9n
White River of Indiana: flatboating cattle down, 11
Whitehead, Chris: imported Shorthorns, 88
Wilder, D.: introduced Patton cattle to Miami Valley, 27
Wilderness Road: as cattle trail, 105
Wilhoit, Thomas: bred Shorthorns, 96
Winchester, Kentucky: cattle estates northwest of, 44-45; S. Clay at court day in, 45; herds assembled at, 113
Winchester, Virginia: drive to area of, 104; large droves passed through, 115; mentioned, 106
Wooster, Ohio: on drover route, 109
Worthington, Thomas: bought stockers from Chickasaws, 14

Xenia, Ohio: mentioned, 59

Yankees: in Ohio Valley, 2
York, Pennsylvania: stall-feeding around, 103
Youatt, William: commented on relation of soils to breeding, 13; theories on breeding, 24
Young Mary: half interest in bought by S. Van Meter and I. Cunningham, 78; bought by J. Harness, 78
Young Phyllis: dropped, 77

Zane's Trace: as pioneer route, 2
Zanesville, Ohio: on pioneer route, 1; on cattle route, 106, 109, 110

www.ingramcontent.com/pod-product-compliance
Lightning Source LLC
Chambersburg PA
CBHW022101160426
43198CB00008B/309